CSIR-UGC NET (JRF) MATHEMATICAL SCIENCE

Topology

Book Series P-11/24

(A Comprehensive Guide for better understanding in Mathematics Students of Various Universities and Colleges in India and Abroad)

By

Dr. Jitendra Singh
Associate Professor
Department of Mathematics
Raj Rishi College, Alwar (Raj.), India, 301001

ISBN: 979-83-431-3705-7
Amazon Kindle Direct Publishing, Columbia, SC USA
https://amazon.com
Copyright © 2025 Dr. Jitendra Singh, All rights reserved

Topology

All rights reserved. No part of this book may be reproduced, stored, or transmitted in any form or by any means—electronic, mechanical, photocopying, recording, or otherwise—without prior written permission from the publisher or author.

First Edition: 2024
Second Edition: 2025

ISBN: 979-83-431-3705-7

Main Distributor:
Sushil Pustak Bhandar, Manni Ka Bad, Alwar (Raj.), India
Contact: +91-9079889132

Printing: Sheetal Offset Jaipur
Latex Typing Setting By: Preety Verma

Published by:
Preety Verma
Editor
Ambedkar Nagar, Alwar (Rajasthan), India
Email: editormathbook193@gmail.com
Website: https://tinyurl.com/drjsrrcalwar

Author Disclaimer

The book, Topology, is for educational purposes only. While efforts have been made to ensure accuracy, the author does not guarantee its completeness or suitability for all situations. Examples are illustrative and not tied to actual systems or projects. Readers should adapt the content responsibly. The author and publisher disclaim liability for errors, omissions, or any resulting damages.

> **Topology**
> **SYLLABUS**
> **Book Series P-11/24**

Topology:

Topology: basis, dense sets, subspace and product topology, separation axioms, connectedness and compactness.

Remarks

The NET-JRF Mathematical Sciences syllabus is comprehensively covered in 24 books, offering topic-wise explanations and detailed solutions. This series is an essential resource for CSIR-UGC NET (JRF), SET, GATE, IIT JAM, Ph.D. entrances, and other competitive exams for Assistant Professor positions and mathematics-based university exams in India. It ensures a strong grasp of core concepts and effective problem-solving skills. Specialized MCQ Series for Mathematics – Designed as per the 2024 New Syllabus of CSIR-UGC NET (JRF) Mathematical Science.

Our Other Books

S.N	Book Title	Book Series
1	Set Theory & the Real Number System	P-1/24
2	Sequences, Series, and Convergence	P-2/24
3	Continuity, Differentiability, and Mean Value Theorems	P-3/24
4	Integration and Improper Integrals	P-4/24
5	Advanced Analysis and Measure	P-5/24
6	Linear Algebra	P-6/24
7	Metric Spaces and Functional Spaces	P-7/24
8	Complex Analysis	P-8/24
9	Combinatorics and Number Theory	P-9/24
10	Abstract Algebra	P-10/24
11	Topology	P-11/24
12	Introduction to Ordinary Differential Equations (ODEs)	P-12/24
13	Advanced ODE and Boundary Value Problems	P-13/24
14	Partial Differential Equations (PDEs)	P-14/24
15	Numerical Analysis	P-15/24
16	Calculus of Variations	P-16/24
17	Integral Equations	P-17/24
18	Classical Mechanics and Advanced Applications	P-18/24
19	Foundations of Probability and Statistics	P-19/24
20	Probability Convergence and Limit Theorems	P-20/24
21	Statistical Inference	P-21/24
22	Advanced Statistical Techniques	P-22/24
24	Operations Research and Queuing Theory	P-24/24

Contents

Preface   8

1 Basic Topology   9
- 1.1 Topological Spaces 9
  - 1.1.1 Definition of a Topological Space 9
  - 1.1.2 Examples of Topological Spaces 9
  - 1.1.3 Basic Terminology and Concepts 10
- 1.2 Basis for a Topology 10
  - 1.2.1 Definition of a Basis 10
  - 1.2.2 Examples of Bases 11
  - 1.2.3 Subbasis for a Topology 11
  - 1.2.4 Examples of Subbases 11
  - 1.2.5 Comparing Topologies via Bases 12
  - 1.2.6 Examples of Topology Comparison 12
- 1.3 Dense Sets and Applications 13
  - 1.3.1 Definition of Dense Sets 13
  - 1.3.2 Applications of Dense Sets 14
  - 1.3.3 Remarks and Further Notes 14
- 1.4 Closed Sets and Their Properties 15
  - 1.4.1 Definition of Closed Sets 15
  - 1.4.2 Examples of Closed Sets 15
  - 1.4.3 Properties of Closed Sets 15
  - 1.4.4 Examples of Closed Sets in Metric Spaces 17
- 1.5 Interior, Closure, and Boundary 18
  - 1.5.1 Interior of a Set 18
  - 1.5.2 Boundary of a Set 18
- 1.6 Continuity and Homeomorphisms 20
  - 1.6.1 Continuous Functions 20

	1.6.2	Homeomorphisms	21
1.7		Compactness and Connectedness	23
	1.7.1	Compactness	23
	1.7.2	Connectedness	24
	1.7.3	Separation Axioms	25
	1.7.4	Normal Spaces (T_4)	25
	1.7.5	Other Separation Axioms	26
1.8		MCQs: Basic Topology	26

2 Subspace and Product Topology — 64
- 2.1 Subspace Topology — 64
  - 2.1.1 Definition — 64
  - 2.1.2 Examples of Subspace Topologies — 64
  - 2.1.3 Basic Properties of Subspace Topology — 65
  - 2.1.4 Generalization of Subspace Topology — 66
  - 2.1.5 Further Examples of Subspace Topologies — 66
- 2.2 Product Topology — 67
  - 2.2.1 Definition — 67
  - 2.2.2 Basic Properties of Product Topology — 68
  - 2.2.3 Further Examples of Product Topologies — 69
  - 2.2.4 Tychonoff's Theorem — 69
- 2.3 Additional Basic Concepts — 69
  - 2.3.1 Closed Sets in Subspace Topology — 69
  - 2.3.2 Continuity in Subspace and Product Topology — 70
  - 2.3.3 Compactness and Hausdorff Spaces — 70
- 2.4 Further Concepts in Subspace and Product Topology — 71
  - 2.4.1 Open Sets in Subspace Topology — 71
  - 2.4.2 Product of Subspaces — 71
  - 2.4.3 Comparison of Subspace Topology and Product Topology — 71
  - 2.4.4 Projections in Product Topology — 72
  - 2.4.5 Subspace and Product Topology on Infinite Products — 72
- 2.5 MCQs: Subspace and Product Topology — 73

3 Separation Axioms — 105
- 3.1 T_0 Spaces (Kolmogorov Spaces) — 105
- 3.2 T_1 Spaces (Frechet Spaces) — 107
- 3.3 T_2 Spaces (Hausdorff Spaces) — 109

3.4	Other Separation Properties	112
	3.4.1 T_3 Spaces (Regular Spaces)	112
	3.4.2 T_4 Spaces (Normal Spaces)	112
	3.4.3 Completely Regular and Tychonoff Spaces	113
3.5	Relations Between Separation Axioms	115
3.6	Additional Separation Properties and Applications	117
	3.6.1 Completely Hausdorff Spaces	117
	3.6.2 Completely Regular Spaces (Tychonoff Spaces)	117
	3.6.3 Applications of Separation Axioms	117
	3.6.4 Connections Between Separation Axioms and Other Topological Properties	119

4 Connectedness and Compactness 156

4.1	Connected Sets and Their Properties	156
	4.1.1 Definition of Connectedness	156
	4.1.2 Disconnected Sets	158
	4.1.3 Applications and Further Studies of Connectedness	158
4.2	Compact Sets and Compactness Criteria	159
	4.2.1 Definition of Compactness	159
	4.2.2 Characterization in $\mathbb{R}$	160
	4.2.3 Properties of Compact Sets	160
	4.2.4 Applications of Compactness	161
4.3	Basic Concepts Related to Compactness and Connectedness	162
	4.3.1 Bolzano–Weierstrass Theorem	162
	4.3.2 Heine–Borel Theorem	162
	4.3.3 Intermediate Value Theorem	163
	4.3.4 Extreme Value Theorem	163
4.4	Additional Concepts Related to Compactness and Connectedness	164
	4.4.1 Path Connectedness	164
	4.4.2 Locally Compact Spaces	164
	4.4.3 Compactness and Sequential Compactness	165
	4.4.4 Applications of Compactness and Connectedness	165
	4.4.5 Product of Compact and Connected Sets	165
4.5	MCQs: Connectedness and Compactness	166

Preface

This book is tailored for aspirants of the CSIR NET (JRF) in Mathematical Sciences, focusing on topology—a key component of the exam syllabus. It is structured into four comprehensive chapters that cover fundamental topological concepts.

Chapter 1 introduces basic topology, starting with the definition of topological spaces, and explores the basis for a topology. It also explains dense sets and their applications, providing essential tools for analysis.

Chapter 2 covers subspace and product topology, offering definitions and examples of the subspace topology and explaining the construction and properties of the product topology, formed through Cartesian products.

Chapter 3 delves into the separation axioms (T_0, T_1, T_2), which are crucial for distinguishing different types of topological spaces, with a focus on Hausdorff spaces and their importance in analysis and geometry.

Chapter 4 focuses on connectedness and compactness, two critical concepts. It explores connected sets and their role in determining the structure of topological spaces, along with compactness and its applications, particularly in analysis.

This book offers a clear, structured guide with examples and exercises to help students build a solid foundation in topology, ensuring they are well-prepared for the CSIR NET (JRF) exam.

Dr. Jitendra Singh

Chapter 1

Basic Topology

1.1 Topological Spaces

1.1.1 Definition of a Topological Space

Let X be a non-empty set. A **topology** on X is a collection $\mathcal{T} \subseteq \mathcal{P}(X)$ of subsets of X satisfying the following properties:

1. **Inclusion of Extremes:** $\emptyset \in \mathcal{T}$ and $X \in \mathcal{T}$.

2. **Closure under Arbitrary Unions:** If $\{U_i\}_{i \in I} \subseteq \mathcal{T}$, then $\bigcup_{i \in I} U_i \in \mathcal{T}$.

3. **Closure under Finite Intersections:** If $U_1, U_2, \ldots, U_n \in \mathcal{T}$ for some $n \in \mathbb{N}$, then $\bigcap_{i=1}^{n} U_i \in \mathcal{T}$.

The pair $(X, \mathcal{T})$ is called a *topological space*, and the sets in $\mathcal{T}$ are called *open sets*.

Remark: A subset $F \subseteq X$ is called **closed** if its complement $X \setminus F$ is open; that is, if $X \setminus F \in \mathcal{T}$.

1.1.2 Examples of Topological Spaces

1. **Discrete Topology:** Let X be any set. The discrete topology on X is defined as $\mathcal{T} = \mathcal{P}(X)$, the power set of X. In this case, every subset of X is open. This is the finest topology on X.

2. **Trivial (Indiscrete) Topology:** Let X be any set. The trivial topology is given by $\mathcal{T} = \{\emptyset, X\}$, containing only the empty set and the entire set. This is the coarsest topology on X.

3. **Standard Topology on** $\mathbb{R}$: Let $X = \mathbb{R}$. The standard topology is generated by the basis consisting of all open intervals (a, b) where $a < b$, $a, b \in \mathbb{R}$. The open sets in this topology are arbitrary unions of such intervals.

4. **Finite Complement Topology:** Let X be an infinite set. Define
$$\mathcal{T} = \{U \subseteq X \mid X \setminus U \text{ is finite or } U = \emptyset\}.$$
This forms a topology called the *finite complement topology*.

5. **Cofinite Topology on** $\mathbb{N}$: Let $X = \mathbb{N}$. Define $\mathcal{T}$ to consist of all subsets $U \subseteq \mathbb{N}$ such that $\mathbb{N} \setminus U$ is finite, along with the empty set. This is a special case of the finite complement topology on a countable set.

1.1.3 Basic Terminology and Concepts

- A set $U \subseteq X$ is **open** if $U \in \mathcal{T}$.
- A set $F \subseteq X$ is **closed** if $X \setminus F \in \mathcal{T}$.
- A **basis** $\mathcal{B}$ for a topology on X is a collection of subsets of X such that every open set can be expressed as a union of elements of $\mathcal{B}$.
- The **topology generated by a basis** $\mathcal{B}$ is the collection of all unions of basis elements.

1.2 Basis for a Topology

1.2.1 Definition of a Basis

Let X be a non-empty set. A collection $\mathcal{B}$ of subsets of X is called a **basis for a topology on** X if:

1. For every $x \in X$, there exists at least one $B \in \mathcal{B}$ such that $x \in B$.
2. If $x \in B_1 \cap B_2$ for some $B_1, B_2 \in \mathcal{B}$, then there exists $B_3 \in \mathcal{B}$ such that $x \in B_3 \subseteq B_1 \cap B_2$.

Given such a collection $\mathcal{B}$, the **topology generated by** $\mathcal{B}$ is:
$$\mathcal{T} = \{U \subseteq X \mid \forall x \in U, \exists B \in \mathcal{B} \text{ such that } x \in B \subseteq U\}.$$

1.2.2 Examples of Bases

1. **Standard Topology on $\mathbb{R}$:**

   $$\mathcal{B} = \{(a,b) \mid a < b,\ a, b \in \mathbb{R}\}$$

   This collection of open intervals generates the usual (Euclidean) topology on $\mathbb{R}$.

2. **Lower Limit Topology (Sorgenfrey Line):**

   $$\mathcal{B} = \{[a,b) \mid a < b\}$$

   This basis generates the lower limit topology, which has more open sets than the standard topology.

3. **Discrete Topology:**

   $$\mathcal{B} = \{\{x\} \mid x \in X\}$$

   This basis generates the discrete topology where every subset of X is open.

1.2.3 Subbasis for a Topology

A collection $\mathcal{S}$ of subsets of X is a **subbasis** for a topology on X if the finite intersections of elements of $\mathcal{S}$ form a basis. That is,

$$\mathcal{B} = \left\{ \bigcap_{i=1}^{n} S_i \ \middle|\ S_i \in \mathcal{S},\ n \in \mathbb{N} \right\}$$

is a basis, and the topology generated by $\mathcal{S}$ is the one generated by $\mathcal{B}$.

1.2.4 Examples of Subbases

1. **Standard Topology on $\mathbb{R}$:**

   $$\mathcal{S} = \{(-\infty, a) \mid a \in \mathbb{R}\} \cup \{(b, \infty) \mid b \in \mathbb{R}\}$$

   The finite intersections produce open intervals (a, b), forming a basis for the standard topology.

2. **Zariski Topology on** $\mathbb{A}^1(\mathbb{C})$: In algebraic geometry, take $X = \mathbb{C}$. Let
$$\mathcal{S} = \{\mathbb{C} \setminus \{z\} \mid z \in \mathbb{C}\}$$
The topology generated by this subbasis is the Zariski topology on $\mathbb{C}$.

3. **Product Topology:** For $X \times Y$, if $\mathcal{S}_X$ and $\mathcal{S}_Y$ are subbases for topologies on X and Y, then
$$\mathcal{S} = \{U \times Y \mid U \in \mathcal{S}_X\} \cup \{X \times V \mid V \in \mathcal{S}_Y\}$$
is a subbasis for the product topology.

1.2.5 Comparing Topologies via Bases

Let $\mathcal{B}_1$ and $\mathcal{B}_2$ be bases for topologies $\mathcal{T}_1$ and $\mathcal{T}_2$ on X, respectively. Then:

$$\mathcal{T}_1 \subseteq \mathcal{T}_2 \iff \forall B_1 \in \mathcal{B}_1, \forall x \in B_1, \exists B_2 \in \mathcal{B}_2 \text{ such that } x \in B_2 \subseteq B_1.$$

1.2.6 Examples of Topology Comparison

1. **Standard vs. Lower Limit Topology on** $\mathbb{R}$: Every open interval (a, b) can be written as a union of half-open intervals $[x, y)$, so:
$$\mathcal{T}_{\text{standard}} \subsetneq \mathcal{T}_{\text{lower limit}}$$

2. **Trivial vs. Discrete Topology:** The trivial topology $\{\emptyset, X\}$ is coarser than any other topology on X, so:
$$\mathcal{T}_{\text{trivial}} \subseteq \mathcal{T}_{\text{discrete}}$$

3. **Metric Topology vs. Zariski Topology on** $\mathbb{C}$: The Zariski topology has very few open sets compared to the standard Euclidean topology:
$$\mathcal{T}_{\text{Zariski}} \subsetneq \mathcal{T}_{\text{Euclidean}}$$

Conclusion and Remarks

- Bases allow us to define topologies with minimal data, by describing only essential open sets.

- Subbases simplify construction by allowing finite intersections to define a full basis.

- Comparing topologies via their bases helps us understand relative coarseness and fineness—an essential aspect in analysis and topology.

1.3 Dense Sets and Applications

1.3.1 Definition of Dense Sets

Let $(X, \mathcal{T})$ be a topological space. A subset $A \subseteq X$ is said to be **dense** in X if every point $x \in X$ either belongs to A or is a limit point of A. Equivalently, A is dense in X if the closure of A equals the whole space:

$$\overline{A} = X.$$

Another equivalent condition is: A is dense in X if every non-empty open subset U of X intersects A, i.e.,

$$\forall U \in \mathcal{T},\ U \neq \emptyset \Rightarrow U \cap A \neq \emptyset.$$

Examples of Dense Sets

1. **Rational Numbers** $\mathbb{Q}$ **in** $\mathbb{R}$: Under the standard topology, the rational numbers are dense in $\mathbb{R}$ because between any two real numbers, there exists a rational number. Hence,

$$\overline{\mathbb{Q}} = \mathbb{R}.$$

2. **Irrational Numbers** $\mathbb{R} \setminus \mathbb{Q}$ **in** $\mathbb{R}$: Likewise, irrational numbers are also dense in $\mathbb{R}$. Every open interval in $\mathbb{R}$ contains both rational and irrational numbers.

3. **Polynomials with Rational Coefficients in** $C[0,1]$: In the space of continuous functions $C[0,1]$ under the sup norm, the set of all polynomials with rational coefficients is dense. This is a consequence of the Weierstrass Approximation Theorem.

More Examples in Diverse Spaces

1. **Trigonometric Polynomials in** $L^2([-\pi, \pi])$:
   The span of $\{\sin(nx), \cos(nx)\}$, for $n \in \mathbb{N}$, is dense in the Hilbert space $L^2([-\pi, \pi])$. This forms the foundation for Fourier analysis.

2. **Step Functions in** $L^1([0,1])$: The collection of step functions (i.e., piecewise constant functions) with rational values is dense in the space of integrable functions $L^1([0,1])$.

3. **Finitely Nonzero Sequences in** ℓ^2: In the Hilbert space ℓ^2 (square-summable sequences), the set of sequences with only finitely many non-zero rational entries is dense.

1.3.2 Applications of Dense Sets

1. **Approximation:** Dense subsets allow for approximation of general elements using simpler or well-understood elements.

   (a) Any continuous function on $[0, 1]$ can be uniformly approximated by polynomials (Weierstrass Theorem).

   (b) Any element in ℓ^2 can be approximated by truncating to finitely many terms.

   (c) In $L^2([0, 1])$, any function can be approximated in norm by piecewise linear or step functions.

2. **Functional Extension and Uniqueness:** If two continuous functions agree on a dense subset, they agree on the whole space.

   (a) If $f, g : X \to Y$ are continuous and $f(x) = g(x)$ for all $x \in A$ where A is dense in X, then $f = g$ on X.

   (b) The identity function is determined by its values on a dense subset.

   (c) Any continuous extension from a dense subset is unique, if it exists.

3. **Separable Spaces:** A space is separable if it has a countable dense subset.

   (a) $\mathbb{R}$ is separable: $\mathbb{Q}$ is a countable dense subset.

   (b) The Hilbert space ℓ^2 is separable: sequences with finitely many rational entries form a countable dense set.

   (c) The space $C[0, 1]$ is separable under the sup norm.

1.3.3 Remarks and Further Notes

- A dense set may not be open or closed.
- In a Hausdorff space, the closure of a singleton is the singleton itself.
- If A is dense in X and $f : X \to Y$ is continuous, then under some conditions, $f(A)$ is dense in $f(X)$.
- Baire's Theorem: In a complete metric space, the countable intersection of open dense sets is still dense.

Chapter 1: Basic Topology

- A subset can be dense in one topology and not in another. For example, $\mathbb{Q}$ is dense in $\mathbb{R}$ with the usual topology, but not under the discrete topology.

1.4 Closed Sets and Their Properties

1.4.1 Definition of Closed Sets

A subset $F \subseteq X$ is called *closed* if its complement $X \setminus F$ is open, i.e., $X \setminus F \in \mathcal{T}$. In other words, a set F is closed if:

$$F \text{ is closed} \iff X \setminus F \text{ is open}.$$

Equivalently, F is closed if it contains all its limit points. This means that if a sequence (or more generally, a net) converges to a point x in F, then x must belong to F. In formal terms, F is closed if for every convergent sequence $\{x_n\}$ in X, if $\lim_{n \to \infty} x_n = x$ and $x_n \in F$ for all n, then $x \in F$.

1.4.2 Examples of Closed Sets

1. **Closed Interval in $\mathbb{R}$:** The closed interval $[a, b] \subseteq \mathbb{R}$ is closed in the standard topology. The complement of $[a, b]$ is the union of two open intervals, $(-\infty, a)$ and (b, ∞), both of which are open in the standard topology.

2. **Empty Set and Whole Space:** Both the empty set $\emptyset$ and the entire space X are closed. This is because their complements (the entire space and the empty set, respectively) are open.

3. **Closed Sets in a Metric Space:** In a metric space (X, d), a set F is closed if it contains all its limit points. For example, the set of integers $\mathbb{Z}$ is closed in $\mathbb{R}$ with the standard topology because the complement, $\mathbb{R} \setminus \mathbb{Z}$, is open.

1.4.3 Properties of Closed Sets

1. **The Whole Space X and the Empty Set $\emptyset$ are Closed:** By definition, both X and $\emptyset$ are closed sets. This is because the complement of X is the empty set, and the complement of $\emptyset$ is X, both of which are open. Thus:

$$X \text{ and } \emptyset \text{ are closed sets.}$$

2. **The Intersection of Any Collection of Closed Sets is Closed:** If $\{F_i\}_{i \in I}$ is a family of closed sets in a topological space X, then their intersection:

$$F = \bigcap_{i \in I} F_i$$

is also closed. This follows from De Morgan's law, which states that:

$$X \setminus \left(\bigcap_{i \in I} F_i \right) = \bigcup_{i \in I} (X \setminus F_i),$$

and since each $X \setminus F_i$ is open, their union is open. Hence, the complement of F is open, implying F is closed.

3. **The Union of a Finite Number of Closed Sets is Closed:**
If $F_1, F_2, \ldots, F_n$ are closed sets, then their union:

$$F = F_1 \cup F_2 \cup \cdots \cup F_n$$

is also closed. This is because the complement of F is the intersection of the complements of the F_i's:

$$X \setminus F = X \setminus (F_1 \cup F_2 \cup \cdots \cup F_n) = (X \setminus F_1) \cap (X \setminus F_2) \cap \cdots \cap (X \setminus F_n).$$

Since the intersection of open sets is open, $X \setminus F$ is open, so F is closed.

4. **Closed Sets are Closed Under Finite Intersections:** While the union of a finite number of closed sets is closed, their intersection is also closed. This follows from the fact that the intersection of two closed sets is closed, and this property extends to finite collections. If $F_1, F_2, \ldots, F_n$ are closed, then:

$$F = F_1 \cap F_2 \cap \cdots \cap F_n$$

is closed. This can be shown using De Morgan's law, which asserts that the complement of an intersection is the union of the complements, and since the complement of each F_i is open, their union is open, so F is closed.

Chapter 1: Basic Topology 17

5. **Closed Sets are Closed Under Countable Intersections in Some Spaces:** In some topological spaces, such as complete metric spaces, the intersection of countably many closed sets may not be closed. However, in spaces like $\mathbb{R}$ (with the standard topology), the intersection of any countable family of closed sets is also closed. This is a crucial property in the study of completeness and compactness.

6. **Closed Sets Contain All Their Limit Points:** A closed set F contains all its limit points. This means that if a sequence $\{x_n\}$ in F converges to a point x, then x must lie in F. For example, the set of rational numbers $\mathbb{Q}$ is not closed in $\mathbb{R}$ because there are sequences of rationals that converge to irrational numbers, which are not in $\mathbb{Q}$.

1.4.4 Examples of Closed Sets in Metric Spaces

1. **Closed Ball in $\mathbb{R}^n$:** A closed ball in $\mathbb{R}^n$ centered at a point $x_0 \in \mathbb{R}^n$ with radius $r > 0$ is closed. This set is defined as:

$$B[x_0, r] = \{x \in \mathbb{R}^n \mid \|x - x_0\| \leq r\}.$$

   The complement of this set is open, and hence the closed ball is closed.

2. **The Set of Limit Points:** In a topological space, the closure of a set A, denoted by $\overline{A}$, is the smallest closed set containing A, and it contains all the limit points of A. The closure of A is closed by definition.

3. **Closed Sets in the Discrete Topology:** In the discrete topology, every set is closed because the complement of any set is open. Thus, for any subset $F \subseteq X$, F is closed in the discrete topology.

Conclusion

Closed sets play a fundamental role in topology, particularly in the study of continuity, convergence, and compactness. Understanding their properties, such as closure under intersections and finite unions, as well as their relationship with limit points, is crucial for many topological concepts.

1.5 Interior, Closure, and Boundary

1.5.1 Interior of a Set

The *interior* of a set $A \subseteq X$, denoted by $\text{int}(A)$, is the largest open set contained in A. More formally,

$$\text{int}(A) = \bigcup \{U \subseteq A \mid U \text{ is open in } X\}.$$

This is the set of all points in A for which there exists an open neighborhood entirely contained within A. If a point $x \in A$ has any open neighborhood that is completely inside A, then x belongs to the interior of A. Thus, the interior of A is the largest open set that is a subset of A.

Remark

The *closure* of a set $A \subseteq X$, denoted by $\overline{A}$, is the smallest closed set containing A. Equivalently, $\overline{A}$ can be defined as the union of A with its limit points. More formally:

$$\overline{A} = A \cup \{x \in X \mid \forall U \text{ open}, x \in U \Rightarrow U \cap A \neq \emptyset\}.$$

The closure of A contains all points of A as well as any limit points of A. A limit point of A is a point where every neighborhood of it contains at least one point from A (which may or may not be in A itself).

1.5.2 Boundary of a Set

The *boundary* of a set $A \subseteq X$, denoted by ∂A, is the set of points that are neither in the interior of A nor in the interior of its complement. In other words,

$$\partial A = \overline{A} \cap \overline{X \setminus A}.$$

Alternatively, the boundary of A can be seen as the set of points where every neighborhood intersects both A and its complement $X \setminus A$. Therefore, boundary points are those that are "on the edge" of A.

Examples

1. **Example 1: For the set $A = (0,1) \subset \mathbb{R}$ in the standard topology:**

   $$\text{int}(A) = (0,1), \quad \overline{A} = [0,1], \quad \partial A = \{0,1\}.$$

   In this case, the interior of A is the set $(0,1)$ itself, as every point in $(0,1)$ has a neighborhood entirely contained in A. The closure of A is the closed interval $[0, 1]$, which includes all the limit points of A (i.e., 0 and 1). The boundary of A consists of the points 0 and 1, which are the boundary points where every neighborhood intersects both A and its complement.

2. **Example 2:** For the set $B = \mathbb{Q} \cap (0, 1)$ in $\mathbb{R}$:

$$\text{int}(B) = \emptyset, \quad \overline{B} = [0, 1], \quad \partial B = [0, 1].$$

In this case, B is the set of rational numbers between 0 and 1. The interior of B is empty because no point in B has a neighborhood entirely contained within B (since between any two rationals, there are irrational numbers). The closure of B is the closed interval $[0, 1]$ because every point in $[0, 1]$ is a limit point of B. The boundary of B is $[0, 1]$, as every point in this interval is either in B or is a limit point of B.

3. **Example 3:** For $C = \mathbb{R}$:

$$\text{int}(C) = \mathbb{R}, \quad \overline{C} = \mathbb{R}, \quad \partial C = \emptyset.$$

The interior of $\mathbb{R}$ is $\mathbb{R}$ itself because every point in $\mathbb{R}$ has a neighborhood entirely contained in $\mathbb{R}$. The closure of $\mathbb{R}$ is also $\mathbb{R}$, as there are no additional limit points outside of $\mathbb{R}$. The boundary of $\mathbb{R}$ is empty because there are no points in $\mathbb{R}$ that are neither in the interior of $\mathbb{R}$ nor in the interior of its complement (which is empty).

4. **Example 4:** For the set $D = [0, 1] \cup [2, 3] \subseteq \mathbb{R}$:

$$\text{int}(D) = (0, 1) \cup (2, 3), \quad \overline{D} = [0, 1] \cup [2, 3], \quad \partial D = \{1, 2\}.$$

In this case, the interior of D is the union of the open intervals $(0, 1)$ and $(2, 3)$, as these are the largest open sets contained in D. The closure of D is $[0, 1] \cup [2, 3]$, which is the smallest closed set containing D. The boundary of D consists of the points $\{1, 2\}$, which are the points that are the limit points of both $[0, 1]$ and $[2, 3]$.

5. **Example 5:** For the set $E = \mathbb{Q}$ in $\mathbb{R}$ (the set of all rational numbers):

$$\text{int}(E) = \emptyset, \quad \overline{E} = \mathbb{R}, \quad \partial E = \mathbb{R}.$$

The interior of $\mathbb{Q}$ is empty because no open interval in $\mathbb{R}$ can be completely contained within $\mathbb{Q}$ (since there are irrational numbers between any two rationals). The closure of $\mathbb{Q}$ is $\mathbb{R}$ because every point in $\mathbb{R}$ is either a rational number or a limit point of rational numbers. The boundary of $\mathbb{Q}$ is the entire real line $\mathbb{R}$ because every point in $\mathbb{R}$ is either in $\mathbb{Q}$ or a limit point of $\mathbb{Q}$.

Properties of Interior, Closure, and Boundary

1. **Interior is an Open Set:** The interior of any set A is always an open set. That is, $\text{int}(A) \in \mathcal{T}$.

2. **Closure is a Closed Set:** The closure of any set A is always a closed set. That is, $\overline{A} \in \mathcal{T}$.

3. **Boundary is a Closed Set:** The boundary of any set A is always a closed set. This is because $\partial A = \overline{A} \cap \overline{X \setminus A}$, and both $\overline{A}$ and $\overline{X \setminus A}$ are closed sets.

4. **Interior of the Closure:** The interior of the closure of a set A is the interior of A:
$$\text{int}(\overline{A}) = \text{int}(A).$$

5. **Closure of the Interior:** The closure of the interior of a set A is always contained in the closure of A:
$$\overline{\text{int}(A)} \subseteq \overline{A}.$$

6. **Boundary of a Set and its Complement:** The boundary of A is the same as the boundary of its complement:
$$\partial A = \partial(X \setminus A).$$

1.6 Continuity and Homeomorphisms

1.6.1 Continuous Functions

A function $f : X \to Y$ between two topological spaces $(X, \mathcal{T}_X)$ and $(Y, \mathcal{T}_Y)$ is called *continuous* if for every open set $V \subseteq Y$, the preimage $f^{-1}(V)$ is open in X. Formally,

$$f \text{ is continuous} \iff \forall V \in \mathcal{T}_Y,\ f^{-1}(V) \in \mathcal{T}_X.$$

Chapter 1: Basic Topology

This definition can also be understood as follows: a function is continuous if the inverse image of every open set in the target space is an open set in the source space. Intuitively, a continuous function does not "break" or "tear" the space, meaning it preserves the topological structure of the spaces involved. Equivalently, f is continuous if for every closed set $C \subseteq Y$, the preimage $f^{-1}(C)$ is closed in X. Since the complement of an open set is closed, these two conditions are equivalent.

1.6.2 Homeomorphisms

A function $f : X \to Y$ is called a *homeomorphism* if it is a continuous bijection, and its inverse $f^{-1} : Y \to X$ is also continuous. In other words, f is a homeomorphism if:

1. f is continuous.

2. f is a bijection (i.e., both injective and surjective).

3. f^{-1} is continuous.

Two spaces X and Y are said to be *homeomorphic* if there exists a homeomorphism between them. Homeomorphic spaces are considered topologically equivalent because they have the same topological structure; one can be "bent" or "stretched" into the other without breaking or gluing.

In topological terms, homeomorphisms preserve all the topological properties of the spaces involved. This means, for example, that two homeomorphic spaces must have the same number of connected components, compactness, and other topological features.

Examples

Example 1: The map $f : (0, 1) \to \mathbb{R}$ given by $f(x) = \tan(\pi x - \pi/2)$ is a homeomorphism.

The function $f(x) = \tan(\pi x - \pi/2)$ maps the open interval $(0, 1)$ onto the real line $\mathbb{R}$, and it is a continuous bijection. Moreover, its inverse, $f^{-1}(y) = \frac{1}{\pi}(\tan^{-1}(y) + \pi/2)$, is also continuous, making f a homeomorphism. This illustrates how certain transformations, such as the tangent function, can preserve the topological structure when applied to intervals.

Example 2: The identity map $id : \mathbb{R} \to \mathbb{R}$ is trivially a homeomorphism.

The identity map $id(x) = x$ is a continuous bijection, and its inverse is itself (which is obviously continuous). Hence, the identity map is a homeomorphism, and it shows that $\mathbb{R}$ is homeomorphic to itself. This example emphasizes that the identity function is always a homeomorphism for any topological space.

Example 3: The map $f : S^1 \to \mathbb{R}^2$ given by $f(\theta) = (\cos\theta, \sin\theta)$ is a homeomorphism between the unit circle S^1 and its image in $\mathbb{R}^2$.

The function $f(\theta) = (\cos\theta, \sin\theta)$ is a continuous bijection from the unit circle S^1 in the plane to its image in $\mathbb{R}^2$, which is the set of points on the unit circle. This map is also invertible: its inverse is given by $f^{-1}(x, y) = \theta = \tan^{-1}(y/x)$, where $(x, y) \in S^1$. Both f and f^{-1} are continuous, so this is a homeomorphism.

Example 4: The map $f : \mathbb{R} \to \mathbb{R}$ given by $f(x) = x^3$ is a homeomorphism.

The function $f(x) = x^3$ is continuous and bijective, and its inverse is given by $f^{-1}(y) = y^{1/3}$, which is also continuous. Therefore, f is a homeomorphism. This example shows how polynomial functions can often be homeomorphisms when their domains are appropriately chosen.

Example 5: The map $f : \mathbb{R}^2 \to \mathbb{R}^2$ given by $f(x, y) = (x, y^3)$ is not a homeomorphism.

Although f is continuous and bijective, its inverse is not continuous. Specifically, the inverse function $f^{-1}(x, z) = (x, z^{1/3})$ is not continuous at points where $z = 0$, because the cube root function has a discontinuity at $z = 0$. Thus, f is not a homeomorphism, even though it is continuous and bijective.

Example 6: The map $f : (0, 1) \to (0, 1)$ given by $f(x) = 1 - x$ is a homeomorphism.

This function is a continuous bijection between the open interval $(0, 1)$ and itself. Its inverse is $f^{-1}(y) = 1 - y$, which is also continuous. Hence, this is a homeomorphism, and it shows how simple functions can also be homeomorphisms.

Properties of Homeomorphisms

1. **Homeomorphisms Preserve Topological Properties:** If $f : X \to Y$ is a homeomorphism, then X and Y are topologically equivalent. This means that they share all topological properties, such as connectedness, compactness, and the number of boundary components.

2. **Composition of Homeomorphisms:** The composition of two homeomorphisms is again a homeomorphism. Specifically, if $f : X \to Y$

Chapter 1: Basic Topology

and $g : Y \to Z$ are homeomorphisms, then their composition $g \circ f : X \to Z$ is also a homeomorphism.

3. **Inverse of a Homeomorphism:** The inverse of a homeomorphism is also a homeomorphism. This is part of the definition of a homeomorphism: not only must f be a continuous bijection, but its inverse f^{-1} must also be continuous.

4. **Topological Invariance:** Homeomorphic spaces are indistinguishable from the perspective of topology. They may look very different geometrically, but they have the same topological structure. This is why topological spaces are studied up to homeomorphism.

1.7 Compactness and Connectedness

1.7.1 Compactness

A topological space X is called *compact* if every open cover of X has a finite subcover. More formally:

Definition 1.7.1. *A space X is compact if for every collection of open sets $\{U_i\}_{i \in I}$ such that $X = \bigcup_{i \in I} U_i$, there exists a finite subset $I_0 \subseteq I$ such that $X = \bigcup_{i \in I_0} U_i$.*

This definition is sometimes referred to as the Heine-Borel property in Euclidean spaces, where a set is compact if and only if it is closed and bounded. However, compactness is a more general topological property that does not require the space to be metric.

Examples of Compact Spaces

1. **Compactness of Closed Intervals in $\mathbb{R}$:** The closed interval $[a, b] \subset \mathbb{R}$ is compact. This can be seen using the Heine-Borel theorem, which states that in $\mathbb{R}$, a subset is compact if and only if it is closed and bounded.

2. **The Unit Circle S^1:** The unit circle $S^1 \subset \mathbb{R}^2$, defined as $S^1 = \{(x, y) \in \mathbb{R}^2 \mid x^2 + y^2 = 1\}$, is compact. It is closed and bounded, and it is compact in the Euclidean topology on $\mathbb{R}^2$.

3. **The Product of Compact Spaces:** The product of a finite number of compact spaces is compact. For example, the product space $[0, 1] \times [0, 1] \subset \mathbb{R}^2$ is compact.

Properties of Compact Spaces
1. Every continuous image of a compact space is compact.

2. A subset of a compact space is compact if and only if it is closed.

3. In a compact space, every sequence has a convergent subsequence (sequential compactness).

4. Compactness is preserved under taking closed subspaces.

1.7.2 Connectedness

A space X is *connected* if it cannot be divided into two non-empty disjoint open sets. Formally:

Definition 1.7.2. *A space X is connected if there do not exist disjoint open sets U and V such that $X = U \cup V$, where $U \neq \emptyset$ and $V \neq \emptyset$.*

Connectedness describes a space that is "in one piece," meaning there are no separations or partitions into disjoint subsets that are open.

Examples of Connected Spaces
1. **The Real Line $\mathbb{R}$:** The real line $\mathbb{R}$ is connected. It cannot be partitioned into two non-empty disjoint open sets.

2. **The Interval $(0, 1)$:** The open interval $(0, 1)$ is connected because there are no non-empty open sets that partition it.

3. **The Unit Circle S^1:** The unit circle S^1 in $\mathbb{R}^2$ is connected because it cannot be divided into two disjoint open sets.

Properties of Connected Spaces
1. The continuous image of a connected space is connected.

2. A space X is connected if and only if the only subsets of X that are both open and closed (i.e., clopen sets) are X and $\emptyset$.

3. The union of two connected spaces that intersect non-trivially is connected.

4. The real line $\mathbb{R}$ is the prototype of a connected space: it is connected in the standard topology, and any connected subset of $\mathbb{R}$ is an interval.

Chapter 1: Basic Topology

1.7.3 Separation Axioms

The separation axioms define the degree to which points and sets in a topological space can be separated by open sets.

Hausdorff Space (T_2)

A topological space X is called a *Hausdorff space* (or T_2 space) if for every pair of distinct points $x, y \in X$, there exist disjoint open sets U and V such that $x \in U$ and $y \in V$. This ensures that points can be "separated" by open neighborhoods.

Definition 1.7.3. *A space X is Hausdorff if for every $x, y \in X$, $x \neq y$, there exist open sets U and V such that:*

$$x \in U, \quad y \in V, \quad U \cap V = \emptyset.$$

Examples of Hausdorff Spaces

1. **Euclidean Space $\mathbb{R}^n$**: Euclidean spaces, such as $\mathbb{R}^2$ and $\mathbb{R}^3$, are Hausdorff. Given two distinct points, we can always find disjoint open neighborhoods around them.

2. **Metric Spaces**: Every metric space is Hausdorff. Given two distinct points, we can always find disjoint open balls around them.

3. **The Discrete Topology:** Any space with the discrete topology is Hausdorff because each singleton set is an open set, so we can always separate distinct points with disjoint open sets.

Properties of Hausdorff Spaces

1. Every compact Hausdorff space is normal (i.e., any two disjoint closed sets can be separated by disjoint open sets).

2. In Hausdorff spaces, limits of sequences (if they exist) are unique.

3. Compact subsets of Hausdorff spaces are closed.

1.7.4 Normal Spaces (T_4)

A topological space X is said to be *normal* if for every pair of disjoint closed sets A and B in X, there exist disjoint open sets U and V such that $A \subseteq U$ and $B \subseteq V$.

Examples of Normal Spaces
1. **Euclidean Space $\mathbb{R}^n$:** Every Euclidean space is normal.
2. **Compact Hausdorff Spaces:** Every compact Hausdorff space is normal.
3. **The Real Line $\mathbb{R}$:** The real line is a normal space.

Properties of Normal Spaces
1. Every compact Hausdorff space is normal.
2. Normality implies that any two disjoint closed sets can be separated by disjoint open sets, which is important for the Urysohn Lemma and Tietze Extension Theorem.

1.7.5 Other Separation Axioms
1. T_0 **(Kolmogorov):** A space is T_0 if for any two distinct points, there exists an open set containing one of the points and not the other.
2. T_1 **(Frechet):** A space is T_1 if for any two distinct points, each has a neighborhood that does not contain the other.
3. T_3 **(Regular):** A space is T_3 if it is T_1 and for any point x and closed set C not containing x, there exist disjoint open sets separating x and C.
4. T_4 **(Normal):** A space is T_4 if it is T_1 and any two disjoint closed sets can be separated by disjoint open sets.

1.8 MCQs: Basic Topology
1. A topological space is defined by a pair (X, τ), where X is a set and τ is:

    (a) A subset of X

    (b) A function from X to the real numbers

    (c) A collection of subsets of X that satisfies certain properties

    (d) A single subset of X that contains all open sets

Answer: (c)
Explanation: τ is a collection of subsets of X that satisfies the properties of a topology, including being closed under arbitrary unions and finite intersections.

2. A basis for a topology τ on a set X is:

    (a) A single subset of X that contains all open sets
    (b) A collection of subsets of X such that every open set in τ can be written as a union of elements from this collection
    (c) A set of functions from X to the real numbers
    (d) A set of subsets of X that are all singletons

Answer: (b)
Explanation: A basis for a topology is a collection of subsets such that every open set in the topology can be expressed as a union of these basis elements.

3. A set A in a topological space (X, τ) is said to be dense in X if:

    (a) The closure of A is equal to X
    (b) A is an open set in X
    (c) A is finite
    (d) A is disjoint from every open set in X

Answer: (a)
Explanation: A set A is dense in X if its closure is X, meaning every point in X is either in A or a limit point of A.

4. A topological space (X, τ) is called a Hausdorff space if:

    (a) Every two distinct points in X have disjoint neighborhoods
    (b) Every open set in τ is closed
    (c) X is compact
    (d) X is connected

Answer: (a)
Explanation: In a Hausdorff space, any two distinct points have disjoint neighborhoods, which ensures that points can be separated by open sets.

5. A subset $A \subset X$ is compact if:

    (a) Every open cover of A has a finite subcover
    (b) A is closed and bounded
    (c) A is open in X
    (d) Every sequence in A converges to a point in A

Answer: (a)
Explanation: A set is compact if every open cover has a finite subcover, which is a key property in topology.

6. In a topological space, a subset A is open if:

    (a) Every point in A has a neighborhood contained in A
    (b) A is the complement of a closed set
    (c) A is closed
    (d) A is bounded

Answer: (a)
Explanation: A set A is open if for every point in A, there exists a neighborhood around it that is entirely contained within A.

7. A subset A of a topological space (X, τ) is closed if:

    (a) Its complement $X \setminus A$ is open
    (b) A contains all its limit points
    (c) A is compact
    (d) Both (a) and (b)

Answer: (d)
Explanation: A set is closed if its complement is open or if it contains all its limit points.

8. The interior of a set A in a topological space (X, τ) is:

Chapter 1: Basic Topology

   (a) The largest open set contained in A

   (b) The smallest closed set containing A

   (c) The complement of the closure of $X \setminus A$

   (d) The set of all limit points of A

   Answer: (a)
   Explanation: The interior of A is the largest open set contained within A, defining the "inside" of A in terms of openness.

9. The closure of a set A in a topological space (X, τ) is:

   (a) The smallest closed set containing A

   (b) The largest open set contained in A

   (c) The complement of the interior of $X \setminus A$

   (d) The union of A and its limit points

   Answer: (a,d)
   Explanation: The closure of A is the union of A and all its limit points, which forms the smallest closed set containing A.

10. Two sets A and B are disjoint if:

    (a) $A \cap B = \emptyset$

    (b) $A \cup B = X$

    (c) $A \subset B$

    (d) $A \cap B \neq \emptyset$

    Answer: (a)
    Explanation: Two sets are disjoint if they have no elements in common, meaning their intersection is empty.

11. A connected topological space is one in which:

    (a) It cannot be divided into two non-empty disjoint open subsets

    (b) Every open set is also closed

    (c) It contains no compact subsets

    (d) Every point is isolated

Answer: (a)

Explanation: A space is connected if it cannot be split into two non-empty, disjoint open sets.

12. A basis for a topology on X is a collection of subsets such that:

    (a) Every open set is a finite intersection of these subsets

    (b) Every open set is a union of these subsets

    (c) Every point in X is contained in exactly one of these subsets

    (d) These subsets are all closed

    Answer: (b)
    Explanation: A basis allows every open set to be expressed as a union of basis elements.

13. In a topological space (X, τ), the set of all limit points of a set A is:

    (a) The interior of A

    (b) The closure of A minus A

    (c) The union of all open sets containing A

    (d) The intersection of all closed sets containing A

    Answer: (b)
    Explanation: The limit points of A are included in the closure of A but not necessarily in A itself.

14. The complement of an open set in a topological space is:

    (a) Closed

    (b) Compact

    (c) Dense

    (d) Connected

    Answer: (a)
    Explanation: The complement of an open set is closed in the given topological space.

15. A set $A \subset X$ is locally compact if:

Chapter 1: Basic Topology

(a) Every point has a neighborhood base of compact sets

(b) Every open set is compact

(c) Every closed set is compact

(d) The space X itself is compact

Answer: (a)

Explanation: Local compactness means that every point has a neighborhood that is compact.

16. In a topological space, the set A is nowhere dense if:

    (a) Its closure has an empty interior

    (b) It is not dense in X

    (c) It is closed

    (d) It is open

    Answer: (a)

    Explanation: A set is nowhere dense if its closure does not contain any non-empty open sets.

17. A topological space (X, τ) is second countable if:

    (a) It has a countable basis for the topology

    (b) It is countable

    (c) Every open cover has a countable subcover

    (d) Every subset is countable

    Answer: (a)

    Explanation: Second countability means that there exists a countable basis for the topology on X.

18. In a topological space, the discrete topology on X is:

    (a) The topology where every subset of X is open

    (b) The topology where only the empty set and X are open

    (c) The topology generated by a single basis element

    (d) The topology where all closed sets are finite

Answer: (a)

Explanation: The discrete topology is the topology in which every subset is an open set.

19. A space is metrizable if:

    (a) It can be endowed with a metric that generates its topology

    (b) It is compact

    (c) It is Hausdorff

    (d) It is connected

    Answer: (a)

    Explanation: A space is metrizable if its topology can be derived from a metric.

20. A path-connected space is one in which:

    (a) Any two points can be connected by a continuous path

    (b) Every path is a homeomorphism

    (c) It is both connected and compact

    (d) It has a path between any two open sets

    Answer: (a)

    Explanation: Path-connectedness means that any two points can be joined by a continuous path within the space.

21. The product topology on $X \times Y$ is defined by:

    (a) The basis consisting of all products of open sets in X and Y

    (b) The basis consisting of all closed sets in $X \times Y$

    (c) The coarsest topology such that the projections to X and Y are continuous

    (d) The finest topology that makes $X \times Y$ compact

    Answer: (a)

    Explanation: The product topology is generated by the basis consisting of products of open sets from X and Y.

22. The subspace topology on a subset $A \subset X$ is:

Chapter 1: Basic Topology 33

   (a) The topology induced by the open sets of X intersected with A

   (b) The topology where A is open in X

   (c) The discrete topology on A

   (d) The coarsest topology that makes the inclusion map continuous

   Answer: (a)
   Explanation: The subspace topology on A consists of intersections of A with the open sets of X.

23. In a topological space (X, τ), a set A is open if:

    (a) It is a union of basis elements

    (b) It is a finite intersection of open sets

    (c) It is closed

    (d) Its complement is compact

    Answer: (a)
    Explanation: An open set is defined as a union of basis elements in the topology.

24. The boundary of a set A in a topological space is:

    (a) The closure of A minus the interior of A

    (b) The intersection of all closed sets containing A

    (c) The union of all open sets containing A

    (d) The complement of the interior of A

    Answer: (a)
    Explanation: The boundary of A is the set of points where A meets its complement, or equivalently, the closure of A minus its interior.

25. The closure of a set A can be defined as:

    (a) The union of A and its limit points

    (b) The intersection of all closed sets containing A

    (c) The smallest open set containing A

    (d) The complement of the interior of $X \setminus A$

Answer: (a)

Explanation: The closure of A includes all points of A and all its limit points.

26. The disjoint union of two topological spaces (X, τ_X) and (Y, τ_Y) is:

    (a) The space $X \cup Y$ with a topology where a set is open if and only if it is the union of an open set in X and an open set in Y
    (b) The space $X \times Y$ with the product topology
    (c) The space $X \cup Y$ with the topology induced by the subspace topologies on X and Y
    (d) The space $X \times Y$ with the discrete topology

    Answer: (a)
    Explanation: The disjoint union is a space where the topology is formed from the individual topologies of X and Y on their respective parts.

27. The coarse topology on X is:

    (a) The topology where the only open sets are $\emptyset$ and X
    (b) The topology where all subsets are open
    (c) The finest topology on X
    (d) The topology generated by all singletons of X

    Answer: (a)
    Explanation: The coarse topology is the one where the only open sets are the empty set and the whole space.

28. The fine topology on X is:

    (a) The topology where all subsets of X are open
    (b) The topology where only the empty set and X are open
    (c) The topology generated by a single open set
    (d) The topology with the fewest open sets

    Answer: (a)
    Explanation: The fine topology is the topology where all possible subsets are open, also known as the discrete topology.

Chapter 1: Basic Topology 35

29. The Alexandrov topology on a set X is:

    (a) The finest topology where every intersection of open sets is open
    (b) The coarsest topology where every union of open sets is open
    (c) The topology where every subset is closed
    (d) The topology where every subset is open

    Answer: (a)
    Explanation: The Alexandrov topology is defined by the property that intersections of open sets are open, which is a finer structure compared to other topologies.

30. A continuity between two topological spaces (X, τ_X) and (Y, τ_Y) is defined as:

    (a) A function $f : X \to Y$ where the preimage of every open set in Y is open in X
    (b) A function $f : X \to Y$ where the image of every open set in X is open in Y
    (c) A function $f : X \to Y$ that is injective
    (d) A function $f : X \to Y$ that is surjective

    Answer: (a)
    Explanation: A function is continuous if the preimage of every open set is open in the domain space.

31. A function $f : X \to Y$ between topological spaces is open if:

    (a) The image of every open set in X is open in Y
    (b) The preimage of every open set in Y is open in X
    (c) The function is continuous and surjective
    (d) The function is continuous and injective

    Answer: (a)
    Explanation: An open function maps open sets to open sets.

32. A function $f : X \to Y$ is closed if:

    (a) The image of every closed set in X is closed in Y

(b) The preimage of every closed set in Y is closed in X

(c) The function is continuous and bijective

(d) The function is continuous and open

Answer: (a)
Explanation: A closed function maps closed sets to closed sets.

33. A topological space X is T1 (or Kolmogorov) if:

    (a) For every pair of distinct points, each has a neighborhood not containing the other

    (b) For every pair of distinct points, there is a neighborhood containing one but not the other

    (c) Every subset is closed

    (d) Every subset is open

    Answer: (a)
    Explanation: In a T1 space, distinct points can be separated by neighborhoods, meaning each point can be isolated from the other.

34. A space is T2 (or Hausdorff) if:

    (a) Any two distinct points can be separated by disjoint neighborhoods

    (b) Every point has a neighborhood that is both open and closed

    (c) Every subset is closed

    (d) Every subset is open

    Answer: (a)
    Explanation: In a Hausdorff space, distinct points have disjoint neighborhoods, which allows for separation of points.

35. A regular space is one in which:

    (a) Any point and a closed set not containing it can be separated by disjoint open sets

    (b) Every open set is closed

    (c) Every closed set is open

Chapter 1: Basic Topology

(d) Every point is isolated

Answer: (a)
Explanation: Regularity means that points and closed sets can be separated by disjoint open sets.

36. A normal space is one in which:

    (a) Any two disjoint closed sets can be separated by disjoint open sets

    (b) Any two distinct points can be separated by disjoint open sets

    (c) Every open set is compact

    (d) Every closed set is compact

    Answer: (a)
    Explanation: Normal spaces allow disjoint closed sets to be separated by disjoint open sets, ensuring a strong form of separation.

37. The Urysohn's Lemma applies to:

    (a) Normal spaces, stating that any two disjoint closed sets can be separated by a continuous function

    (b) Compact spaces, stating that every open cover has a finite subcover

    (c) Hausdorff spaces, stating that every two distinct points have disjoint neighborhoods

    (d) Connected spaces, stating that they cannot be divided into disjoint open sets

    Answer: (a)
    Explanation: Urysohn's Lemma is a result in normal spaces about separating disjoint closed sets by continuous functions.

38. The Stone–Čech compactification of a space X is:

    (a) The largest compact Hausdorff space containing X as a dense subset

    (b) The largest open set containing X

(c) The smallest compact space containing X

(d) The smallest Hausdorff space containing X

Answer: (a)
Explanation: The Stone–Čech compactification is a way to extend a space to a compact Hausdorff space while preserving X as a dense subset.

39. In the context of topological groups, a group G with a topology is a topological group if:

    (a) The group operations (multiplication and inversion) are continuous

    (b) The group is compact

    (c) The group is connected

    (d) The group is Hausdorff

Answer: (a)
Explanation: A topological group is one where the group operations are continuous with respect to the topology.

40. A space is locally compact Hausdorff if:

    (a) Every point has a neighborhood base of compact sets and the space is Hausdorff

    (b) Every point has a neighborhood base of open sets

    (c) Every closed set is compact

    (d) Every open set is compact

Answer: (a)
Explanation: Local compactness and Hausdorff property together imply that every point has a compact neighborhood.

41. A topological vector space is a vector space with a topology such that:

    (a) The vector space operations (vector addition and scalar multiplication) are continuous

    (b) The vector space is compact

Chapter 1: Basic Topology

(c) The vector space is connected

(d) The vector space is Hausdorff

Answer: (a)

Explanation: In a topological vector space, both vector addition and scalar multiplication must be continuous operations with respect to the topology.

42. The Zariski topology on an algebraic variety is characterized by:

    (a) Closed sets are defined as the zeros of polynomial functions

    (b) Open sets are defined as the complements of polynomial functions

    (c) The topology where every subset is open

    (d) The topology where every subset is closed

    Answer: (a)

    Explanation: The Zariski topology is defined by its closed sets being algebraic sets, i.e., zero sets of polynomials.

43. The compact-open topology is used in:

    (a) Function spaces, where the topology is generated by compact subsets and open sets

    (b) Metric spaces, where the topology is defined by distances

    (c) Topological vector spaces, where the topology is induced by vector space operations

    (d) Product spaces, where the topology is generated by product bases

    Answer: (a)

    Explanation: The compact-open topology is used to endow spaces of continuous functions with a topology based on compact subsets and open sets.

44. A space is first-countable if:

    (a) Every point has a countable neighborhood base

    (b) Every open set is countable

    (c) Every closed set is countable

(d) Every subset is countable

Answer: (a)
Explanation: First-countability means that every point in the space has a countable collection of neighborhoods that forms a basis for the topology around that point.

45. A topological space X is compact Hausdorff if:

    (a) X is compact and every two distinct points can be separated by disjoint open sets
    (b) X is compact and every open set is closed
    (c) X is Hausdorff and every closed set is compact
    (d) X is connected and every point is isolated

Answer: (a)
Explanation: Compact Hausdorff spaces are both compact and Hausdorff, providing a strong separation property.

46. A locally compact space is one in which:

    (a) Every point has a neighborhood base of compact sets
    (b) Every open set is compact
    (c) Every closed set is compact
    (d) Every subset is compact

Answer: (a)
Explanation: Local compactness means that around every point, you can find neighborhoods that are compact.

47. In a metrizable space:

    (a) There exists a metric that induces the topology
    (b) The space is compact
    (c) The space is connected
    (d) The space is Hausdorff

Chapter 1: Basic Topology

Answer: (a)
Explanation: Metrizability refers to the ability to describe the space with a metric that generates its topology.

48. In a T4 (or normal Hausdorff) space, any two disjoint closed sets:

    (a) Can be separated by disjoint open sets

    (b) Have a common neighborhood

    (c) Can be separated by a continuous function

    (d) Must be compact

    Answer: (a)
    Explanation: In a T4 space, any two disjoint closed sets can be separated by disjoint open sets.

49. A space is compact Hausdorff if:

    (a) Every open cover has a finite subcover and any two distinct points can be separated by disjoint neighborhoods

    (b) Every subset is closed and the space is compact

    (c) The space is connected and every point is isolated

    (d) Every open set is compact

    Answer: (a)
    Explanation: A space is compact Hausdorff if it is both compact and Hausdorff, ensuring that any two distinct points can be separated by disjoint open sets.

50. The Hamel basis of a vector space is:

    (a) A basis where every vector in the space can be written uniquely as a finite linear combination of the basis elements

    (b) A set of vectors that spans the space but is not necessarily linearly independent

    (c) An infinite set of vectors that spans the space

    (d) A set of vectors that is orthogonal and spans the space

Answer: (a)

Explanation: A Hamel basis allows every vector in the space to be written uniquely as a finite linear combination of the basis elements.

51. A space is path-connected if:

    (a) Any two points in the space can be joined by a continuous path

    (b) Any two points can be connected by a sequence of open sets

    (c) Every open set is connected

    (d) Every closed set is connected

    Answer: (a)

    Explanation: Path-connectedness means there exists a continuous path between any two points in the space.

52. The Stone–Čech compactification of X has the following property:

    (a) It is the largest compact Hausdorff space in which X can be embedded densely

    (b) It is the smallest compact space containing X

    (c) It is the largest space where every subset of X is open

    (d) It is the smallest space containing X with a discrete topology

    Answer: (a)

    Explanation: The Stone–Čech compactification provides a way to embed X into a compact Hausdorff space where X is dense.

53. The discrete topology on a set X is characterized by:

    (a) Every subset of X is open

    (b) Only $\emptyset$ and X are open

    (c) The only closed sets are $\emptyset$ and X

    (d) All sets are closed

    Answer: (a)

    Explanation: In the discrete topology, every subset of X is open, making it the finest topology possible on X.

54. The lower semi-continuous function is defined as:

(a) A function where the preimage of every open set is open

(b) A function where the preimage of every closed set is closed

(c) A function where the preimage of every open set is closed

(d) A function where the preimage of every closed set is open

Answer: (d)
Explanation: A lower semi-continuous function is defined by the preimage of every closed set being open.

55. A quasi-compact space is:

    (a) A space where every open cover has a finite subcover

    (b) A space where every closed set is compact

    (c) A space where every open set is compact

    (d) A space where the space itself is compact

    Answer: (a)
    Explanation: A quasi-compact space is one in which every open cover has a finite subcover.

56. In the Kuratowski closure-complement problem, the maximum number of distinct sets you can form from a given set X using closure and complement operations is:

    (a) 16

    (b) 8

    (c) 4

    (d) 2

    Answer: (a)
    Explanation: The Kuratowski closure-complement problem states that there can be at most 16 distinct sets formed using closure and complement operations.

57. The countable compactness of a space refers to:

    (a) Every countable open cover has a finite subcover

    (b) Every subset is countable

(c) The space is compact and countable

(d) Every closed subset is countable

Answer: (a)
Explanation: Countable compactness means every countable open cover has a finite subcover.

58. A locally connected space is:

    (a) A space where every point has a neighborhood base consisting of connected sets

    (b) A space where every connected set is open

    (c) A space where every subset is connected

    (d) A space where every open set is connected

    Answer: (a)
    Explanation: Local connectedness means every point has a neighborhood base made up of connected sets.

59. A regular Hausdorff space is:

    (a) A space where every point and closed set not containing it can be separated by disjoint open sets and any two distinct points can be separated by disjoint neighborhoods

    (b) A space where every closed set is compact

    (c) A space where every open set is closed

    (d) A space where every subset is open

    Answer: (a)
    Explanation: Regular Hausdorff spaces are those where points and closed sets can be separated and points can be separated by disjoint neighborhoods.

60. The Alexandrov compactification of a non-compact space X is:

    (a) The compactification of X where the added point is a limit point of every sequence in X

    (b) The smallest compact space containing X

Chapter 1: Basic Topology

   (c) The largest compact space containing X as a dense subset

   (d) The space X with a discrete topology

   Answer: (a)
   Explanation: The Alexandrov compactification adds a point to X to make it compact, where the new point is a limit point of every sequence in X.

61. A T0 space is also known as:

    (a) Kolmogorov space

    (b) Hausdorff space

    (c) Regular space

    (d) Normal space

    Answer: (a)
    Explanation: A T0 space is also known as a Kolmogorov space, where for any two distinct points, at least one can be separated from the other by an open set.

62. In a compact space, every open cover:

    (a) Has a finite subcover

    (b) Is a finite cover

    (c) Is countable

    (d) Has at least one infinite subcover

    Answer: (a)
    Explanation: Compact spaces have the property that every open cover has a finite subcover.

63. A regular space can be described as:

    (a) A space where every point and a closed set not containing it can be separated by disjoint open sets

    (b) A space where every point has a neighborhood that is both open and closed

    (c) A space where every open set is also closed

(d) A space where every closed set is compact

Answer: (a)
Explanation: Regular spaces allow for separation of points from closed sets not containing them by disjoint open sets.

64. The final topology on a space X with respect to a function f is:

    (a) The finest topology on X that makes f continuous

    (b) The coarsest topology on X that makes f continuous

    (c) The topology where f is a homeomorphism

    (d) The topology where f is an open map

    Answer: (a)
    Explanation: The final topology is the finest topology on X that makes f continuous.

65. The initial topology on X with respect to a collection of functions $\{f_i\}$ is:

    (a) The coarsest topology on X that makes each f_i continuous

    (b) The finest topology on X that makes each f_i continuous

    (c) The topology where each f_i is an open map

    (d) The topology where each f_i is a closed map

    Answer: (a)
    Explanation: The initial topology is the coarsest topology that makes each function f_i continuous.

66. A space is totally disconnected if:

    (a) The only connected subsets are the singletons

    (b) The space itself is connected

    (c) Every open set is disconnected

    (d) Every closed set is connected

    Answer: (a)
    Explanation: Totally disconnected spaces have the property that the only connected subsets are single points.

Chapter 1: Basic Topology

67. The quotient topology is used to:

    (a) Define a topology on the quotient space $X/\sim$ where $\sim$ is an equivalence relation
    (b) Define a topology on the product space $X \times Y$
    (c) Define a topology on a subspace of X
    (d) Define a topology where every subset is open

    Answer: (a)
    Explanation: The quotient topology is used to define a topology on a quotient space formed by an equivalence relation on X.

68. In a first-countable space:

    (a) Every point has a countable neighborhood base
    (b) Every subset is countable
    (c) Every open set is countable
    (d) Every closed set is countable

    Answer: (a)
    Explanation: First-countability means every point has a countable collection of neighborhoods that forms a basis around that point.

69. In a Hausdorff space, any two distinct points:

    (a) Can be separated by disjoint open sets
    (b) Must have a common neighborhood
    (c) Must be connected
    (d) Must be in the same connected component

    Answer: (a)
    Explanation: Hausdorff spaces have the property that any two distinct points can be separated by disjoint open sets.

70. A disconnected space is one where:

    (a) It can be partitioned into two disjoint nonempty open sets
    (b) Every subset is disconnected

(c) The space itself is connected

(d) Every open set is disconnected

Answer: (a)
Explanation: Disconnected spaces can be divided into two disjoint nonempty open sets.

71. The sublattice of a topological space X is:

    (a) A collection of open sets that forms a lattice under set union and intersection

    (b) A collection of closed sets that forms a lattice under set union and intersection

    (c) A collection of continuous functions that forms a lattice

    (d) A subspace of X with a discrete topology

    Answer: (a)
    Explanation: A sublattice in topology refers to a collection of open sets that is closed under finite union and intersection.

72. The Alexandrov topology on X is characterized by:

    (a) Every intersection of open sets is open

    (b) Every union of open sets is open

    (c) Every set is closed

    (d) Every set is compact

    Answer: (a)
    Explanation: In the Alexandrov topology, arbitrary intersections of open sets are open, which is different from the usual topologies where only finite intersections are guaranteed to be open.

73. A regular space X is one where:

    (a) For any point x and closed set C not containing x, there exist disjoint open sets separating x and C

    (b) For any two distinct points, there exist disjoint open sets separating them

Chapter 1: Basic Topology

(c) Every subset is open

(d) Every open set is compact

Answer: (a)
Explanation: Regular spaces allow for separation of points from closed sets not containing them by disjoint open sets.

74. A topological space is a set X equipped with:

    (a) A collection of subsets called open sets satisfying certain axioms

    (b) A metric defining distances

    (c) A group structure

    (d) A vector space structure

Answer: (a)
Explanation: A topological space is defined by a collection of open sets satisfying the axioms: the empty set and X are open, unions of open sets are open, and finite intersections of open sets are open.

75. A subset $U \subseteq X$ in a topological space X is open if:

    (a) It is in the topology of X

    (b) Its complement is open

    (c) It is closed

    (d) It is finite

Answer: (a)
Explanation: By definition, a set is open if it belongs to the topology of X.

76. A subset $C \subseteq X$ in a topological space is closed if:

    (a) Its complement $X \setminus C$ is open

    (b) It is open

    (c) It is compact

    (d) It is finite

Answer: (a)

Explanation: A set is closed if its complement is open, by the definition of a topological space.

77. The trivial topology on a set X consists of:

    (a) $\{\emptyset, X\}$
    (b) All subsets of X
    (c) Only $\emptyset$
    (d) Only X

    Answer: (a)
    Explanation: The trivial topology includes only the empty set and the entire set X.

78. The discrete topology on a set X consists of:

    (a) All subsets of X
    (b) $\{\emptyset, X\}$
    (c) Only finite subsets
    (d) Only infinite subsets

    Answer: (a)
    Explanation: In the discrete topology, every subset of X is open.

79. A function $f : X \to Y$ between topological spaces is continuous if:

    (a) The preimage of every open set in Y is open in X
    (b) The image of every open set in X is open in Y
    (c) f is bijective
    (d) f is surjective

    Answer: (a)
    Explanation: Continuity is defined by the preimage of open sets being open.

80. A topological space X is Hausdorff if:

    (a) Any two distinct points can be separated by disjoint open sets

(b) It is compact

(c) It is connected

(d) It is finite

Answer: (a)
Explanation: A space is Hausdorff if for any $x \neq y$, there exist disjoint open sets $U \ni x$ and $V \ni y$.

81. Every metric space is:

    (a) Hausdorff

    (b) Non-Hausdorff

    (c) Compact

    (d) Connected

    Answer: (a)
    Explanation: In a metric space, distinct points can be separated by open balls, making it Hausdorff.

82. A subset $K \subseteq X$ is compact if:

    (a) Every open cover of K has a finite subcover

    (b) It is closed

    (c) It is bounded

    (d) It is finite

    Answer: (a)
    Explanation: Compactness is defined by the finite subcover property for open covers.

83. In a Hausdorff space, every compact subset is:

    (a) Closed

    (b) Open

    (c) Dense

    (d) Finite

Answer: (a)

Explanation: Compact subsets in Hausdorff spaces are closed, as points outside can be separated from the subset.

84. The interval $[0, 1] \subset \mathbb{R}$ with the standard topology is:

    (a) Compact

    (b) Not compact

    (c) Open

    (d) Infinite

    Answer: (a)

    Explanation: By the Heine-Borel theorem, $[0, 1]$ is compact as it is closed and bounded in $\mathbb{R}$.

85. A topological space X is connected if:

    (a) It cannot be written as the union of two disjoint non-empty open sets

    (b) It is compact

    (c) It is Hausdorff

    (d) It is finite

    Answer: (a)

    Explanation: A space is connected if it has no non-trivial clopen sets.

86. The real line $\mathbb{R}$ with the standard topology is:

    (a) Connected

    (b) Not connected

    (c) Compact

    (d) Finite

    Answer: (a)

    Explanation: $\mathbb{R}$ is connected, as intervals are the only connected subsets of $\mathbb{R}$.

87. A homeomorphism between topological spaces X and Y is:

Chapter 1: Basic Topology

(a) A continuous bijection with a continuous inverse

(b) A continuous function

(c) A bijective function

(d) An open function

Answer: (a)

Explanation: A homeomorphism is a continuous bijection with a continuous inverse, preserving topological properties.

88. The image of a compact set under a continuous function is:

    (a) Compact
    (b) Open
    (c) Closed
    (d) Dense

Answer: (a)

Explanation: Continuous functions preserve compactness.

89. The image of a connected set under a continuous function is:

    (a) Connected
    (b) Compact
    (c) Open
    (d) Closed

Answer: (a)

Explanation: Continuous functions preserve connectedness.

90. A basis for a topology on X is:

    (a) A collection of subsets whose unions generate all open sets
    (b) A collection of all open sets
    (c) A collection of closed sets
    (d) A collection of compact sets

Answer: (a)

Explanation: A basis generates the topology via unions of its elements.

91. In a metric space (X, d), the open ball $B(x, \epsilon)$ is:

    (a) $\{y \in X \mid d(x, y) < \epsilon\}$
    (b) $\{y \in X \mid d(x, y) \leq \epsilon\}$
    (c) $\{y \in X \mid d(x, y) = \epsilon\}$
    (d) $\{y \in X \mid d(x, y) > \epsilon\}$

    Answer: (a)
    Explanation: An open ball is defined by points within distance ϵ from x.

92. The topology induced by a metric d on a set X has open sets:

    (a) Arbitrary unions of open balls
    (b) Finite unions of open balls
    (c) Intersections of open balls
    (d) Closed balls

    Answer: (a)
    Explanation: The metric topology is generated by unions of open balls.

93. A topological space is separable if:

    (a) It has a countable dense subset
    (b) It is compact
    (c) It is connected
    (d) It is finite

    Answer: (a)
    Explanation: A separable space has a countable subset whose closure is the entire space.

94. The space $\mathbb{R}$ with the standard topology is:

    (a) Separable
    (b) Not separable
    (c) Compact

Chapter 1: Basic Topology

(d) Finite

Answer: (a)
Explanation: $\mathbb{Q}$ is a countable dense subset of $\mathbb{R}$, so it is separable.

95. A subset $D \subseteq X$ is dense if:

    (a) Its closure is X

    (b) It is open

    (c) It is compact

    (d) It is finite

Answer: (a)
Explanation: A set is dense if every point in X is a limit point of D.

96. The closure of a set $A \subseteq X$ is:

    (a) The smallest closed set containing A

    (b) The largest open set contained in A

    (c) The set of all open points in A

    (d) The set A

Answer: (a)
Explanation: The closure is the intersection of all closed sets containing A.

97. A point $x \in X$ is a limit point of a set $A \subseteq X$ if:

    (a) Every neighborhood of x contains a point of A other than x

    (b) $x \in A$

    (c) x is in the interior of A

    (d) x is isolated

Answer: (a)
Explanation: A limit point has points of A arbitrarily close, excluding itself.

98. A topological space is T_1 if:

(a) For any two distinct points, each has an open set not containing the other

(b) It is Hausdorff

(c) It is compact

(d) It is connected

Answer: (a)
Explanation: A T_1 space allows separation of points by open sets, weaker than Hausdorff.

99. Every Hausdorff space is:

    (a) T_1

    (b) Not T_1

    (c) Compact

    (d) Connected

    Answer: (a)
    Explanation: Hausdorff implies T_1, as disjoint open sets provide the required separation.

100. A space is path-connected if:

     (a) Any two points can be joined by a continuous path

     (b) It is connected

     (c) It is compact

     (d) It is Hausdorff

     Answer: (a)
     Explanation: Path-connectedness means there exists a continuous function from $[0, 1]$ joining any two points.

101. Every path-connected space is:

     (a) Connected

     (b) Not connected

     (c) Compact

Chapter 1: Basic Topology 57

(d) Hausdorff

Answer: (a)
Explanation: Path-connectedness implies connectedness, as a disconnection would prevent paths.

102. The product topology on $X \times Y$ has a basis consisting of:

    (a) Sets of the form $U \times V$, where U is open in X and V is open in Y
    (b) Sets of the form $U \times Y$
    (c) Sets of the form $X \times V$
    (d) All subsets of $X \times Y$

    Answer: (a)
    Explanation: The product topology is generated by products of open sets.

103. Tychonoff's theorem states that:

    (a) The product of any collection of compact spaces is compact
    (b) The product of connected spaces is connected
    (c) The product of Hausdorff spaces is Hausdorff
    (d) The product of separable spaces is separable

    Answer: (a)
    Explanation: Tychonoff's theorem guarantees compactness of arbitrary products of compact spaces.

104. A topological space is locally compact if:

    (a) Every point has a compact neighborhood
    (b) The space is compact
    (c) The space is connected
    (d) The space is finite

    Answer: (a)
    Explanation: Local compactness means every point is contained in a compact set's interior.

105. The space $\mathbb{R}$ with the standard topology is:

    (a) Locally compact

    (b) Not locally compact

    (c) Compact

    (d) Finite

    Answer: (a)
    Explanation: Every point in $\mathbb{R}$ has a compact neighborhood, e.g., a closed interval.

106. A continuous function from a compact space to a Hausdorff space is:

    (a) Closed

    (b) Open

    (c) Bijective

    (d) Surjective

    Answer: (a)
    Explanation: Such functions map closed sets to closed sets, as compact sets are closed in Hausdorff spaces.

107. The interior of a set $A \subseteq X$ is:

    (a) The largest open set contained in A

    (b) The smallest closed set containing A

    (c) The set of limit points of A

    (d) The set A

    Answer: (a)
    Explanation: The interior is the union of all open sets contained in A.

108. A space is second-countable if:

    (a) It has a countable basis

    (b) It is separable

    (c) It is compact

Chapter 1: Basic Topology

(d) It is connected

Answer: (a)
Explanation: A second-countable space has a countable collection of open sets generating the topology.

109. The space $\mathbb{R}$ with the standard topology is:

    (a) Second-countable

    (b) Not second-countable

    (c) Compact

    (d) Finite

Answer: (a)
Explanation: The open intervals with rational endpoints form a countable basis for $\mathbb{R}$.

110. A subspace $Y \subseteq X$ inherits the topology where open sets are:

    (a) Intersections of open sets in X with Y

    (b) All open sets in X

    (c) All subsets of Y

    (d) Closed sets in Y

Answer: (a)
Explanation: The subspace topology on Y consists of sets $U \cap Y$, where U is open in X.

111. A set $A \subseteq X$ is clopen if:

    (a) It is both open and closed

    (b) It is neither open nor closed

    (c) It is compact

    (d) It is dense

Answer: (a)
Explanation: A clopen set is both open and closed in the topology.

112. In the discrete topology, every subset is:

(a) Clopen

(b) Open but not closed

(c) Closed but not open

(d) Neither open nor closed

Answer: (a)
Explanation: Every subset is open, hence also closed, in the discrete topology.

113. A topological space is normal if:

    (a) Any two disjoint closed sets can be separated by disjoint open sets

    (b) It is Hausdorff

    (c) It is compact

    (d) It is connected

    Answer: (a)
    Explanation: Normality means disjoint closed sets have disjoint open neighborhoods.

114. Every compact Hausdorff space is:

    (a) Normal

    (b) Not normal

    (c) Connected

    (d) Separable

    Answer: (a)
    Explanation: Compact Hausdorff spaces are normal, as they allow separation of closed sets.

115. The one-point compactification of $\mathbb{R}$ is homeomorphic to:

    (a) The circle S^1

    (b) $\mathbb{R}$

    (c) $[0, 1]$

(d) $\mathbb{R}^2$

Answer: (a)
Explanation: The one-point compactification of $\mathbb{R}$ adds a point at infinity, forming a space homeomorphic to the circle.

116. A function $f : X \to Y$ is open if:

    (a) The image of every open set in X is open in Y

    (b) The preimage of every open set is open

    (c) It is continuous

    (d) It is bijective

    Answer: (a)
    Explanation: An open function maps open sets to open sets.

117. A topological space is completely regular if:

    (a) It is T_1 and points can be separated from closed sets by continuous functions

    (b) It is compact

    (c) It is connected

    (d) It is finite

    Answer: (a)
    Explanation: Complete regularity allows separation of a point and a disjoint closed set by a continuous function to $[0, 1]$.

118. The Cantor set is:

    (a) Compact, totally disconnected, and perfect

    (b) Connected

    (c) Open

    (d) Finite

    Answer: (a)
    Explanation: The Cantor set is compact, has no isolated points (perfect), and no non-trivial connected subsets (totally disconnected).

119. A space is totally disconnected if:

   (a) Its only connected subsets are singletons

   (b) It is not connected

   (c) It is compact

   (d) It is Hausdorff

   Answer: (a)
   Explanation: A totally disconnected space has only trivial connected subsets.

120. The co-finite topology on an infinite set X has open sets:

   (a) $\emptyset$ and sets with finite complements

   (b) All subsets

   (c) Finite subsets

   (d) Infinite subsets

   Answer: (a)
   Explanation: The co-finite topology includes $\emptyset$ and sets whose complements are finite.

121. A topological space is Lindelöf if:

   (a) Every open cover has a countable subcover

   (b) It is compact

   (c) It is connected

   (d) It is finite

   Answer: (a)
   Explanation: A Lindelöf space has a countable subcover for any open cover.

122. The space $\mathbb{R}$ with the standard topology is:

   (a) Lindelöf

   (b) Not Lindelöf

   (c) Compact

(d) Finite

Answer: (a)

Explanation: $\mathbb{R}$ is second-countable, hence Lindelöf, as every open cover has a countable subcover.

Chapter 2

Subspace and Product Topology

2.1 Subspace Topology

2.1.1 Definition

Given a topological space (X, τ) and a subset $Y \subseteq X$, the **subspace topology** on Y is defined as follows:

$$\tau_Y = \{U \cap Y \mid U \in \tau\}.$$

That is, a set $V \subseteq Y$ is open in the subspace topology if and only if there exists an open set $U \subseteq X$ such that $V = U \cap Y$.

In simpler terms, a set in the subspace topology is the intersection of Y with an open set from the original topology of X. This topology on Y is sometimes referred to as the induced or inherited topology from X.

2.1.2 Examples of Subspace Topologies

Example 1: Real Line

Consider the real line $\mathbb{R}$ with the standard topology, and let $Y = [0, 1] \subset \mathbb{R}$. The subspace topology on Y consists of sets of the form $U \cap [0, 1]$ where U is open in $\mathbb{R}$. For example, the open set $(0, 2) \subseteq \mathbb{R}$ intersects with $[0, 1]$ to give $(0, 1)$, which is open in the subspace topology on $[0, 1]$.

Example 2: Unit Circle in $\mathbb{R}^2$

Let $X = \mathbb{R}^2$ with the standard topology, and consider the subspace $Y = \mathbb{S}^1$ (the unit circle). The subspace topology on Y consists of sets of the form $U \cap \mathbb{S}^1$, where U is open in $\mathbb{R}^2$. For example, if $U = B((0, 0), 2)$ is the open

Chapter 2: Subspace and Product Topology

ball of radius 2 around the origin in $\mathbb{R}^2$, the intersection $U \cap \mathbb{S}^1$ will form an open arc on the circle. This open arc is an open set in the subspace topology of $\mathbb{S}^1$.

Example 3: First Quadrant in $\mathbb{R}^2$

Let $X = \mathbb{R}^2$ with the Euclidean topology, and let $Y = \{(x, y) \mid x \geq 0 \text{ and } y \geq 0\}$. The subspace topology on Y will be the topology inherited from $\mathbb{R}^2$, consisting of all sets of the form $U \cap Y$, where U is open in $\mathbb{R}^2$. An example of such a set is the intersection of an open set $U = B((1,1), 1)$, an open ball centered at $(1, 1)$ with radius 1, with the subset Y. This gives a quarter circle shaped region open in the subspace topology on Y.

2.1.3 Basic Properties of Subspace Topology

- **Continuity of the Inclusion Map:** The inclusion map $i : Y \hookrightarrow X$, where $i(y) = y$ for all $y \in Y$, is always continuous. This is because the preimage of any open set in X under i is just the intersection of that open set with Y, which is open in the subspace topology on Y.

- **Hausdorff Property:** If X is Hausdorff, then Y is Hausdorff in the subspace topology. This follows from the fact that if X has the Hausdorff property, then any two distinct points in Y can be separated by open sets in X, which will also separate them in the subspace topology.

- **Compactness:** If X is compact, then Y is compact in the subspace topology. This is a direct consequence of the fact that any open cover of Y (in the subspace topology) can be extended to an open cover of X, which must have a finite subcover by the compactness of X. The intersection of this finite subcover with Y forms a finite subcover of Y.

- **Closed Sets:** If A is closed in X, then $A \cap Y$ is closed in Y in the subspace topology. Conversely, if $A \cap Y$ is closed in Y, then A is closed in X (provided A is closed in X).

- **Openness of Sets:** If U is open in X, then $U \cap Y$ is open in the subspace topology on Y. However, the reverse is not always true: a set that is open in the subspace topology may not be open in the larger space.

2.1.4 Generalization of Subspace Topology

The concept of a subspace topology generalizes to more complex spaces. Suppose X is a topological space, and Y is any subset of X. The subspace topology is one of the simplest methods for "inducing" a topology on a subset, and it is a critical concept in many areas of topology, including the study of continuity and boundary behavior in geometric contexts.

For a more general framework, the subspace topology is crucial in understanding the behavior of subsets in a given topological space. This includes the study of subsets in the product topology, quotient topology, and more specialized constructions such as the Alexandrov topology or upper and lower limit topologies.

Subspace Topology and Continuity

Another important result related to the subspace topology is the following characterization of continuity:

Proposition 2.1.1. *A function $f : X \to Y$ is continuous if and only if for every subset $A \subseteq X$, the preimage $f^{-1}(A)$ is open in the subspace topology on X.*

This is often used in proving the continuity of functions between subspaces and is one of the main reasons the subspace topology is so fundamental.

2.1.5 Further Examples of Subspace Topologies

Example 4: Subspace of a Discrete Space

Let $X = \mathbb{N}$, the set of natural numbers, with the discrete topology, and let $Y = \{1, 2, 3\}$. The subspace topology on Y will also be the discrete topology, because every subset of Y is open in the subspace topology, just as in the discrete topology on X. Therefore, the subspace topology on any subset of a discrete space is always the discrete topology.

Example 5: Subspace of a Compact Space

Consider $X = [0, 1]$ with the standard topology. Let $Y = \{\frac{1}{n} \mid n \in \mathbb{N}\}$ be a subset of X. The subspace topology on Y is the discrete topology because for every point in Y, there exists an open set in X that only contains that point in the subspace Y. Therefore, the subspace topology on Y in this case is discrete, even though X itself is compact.

Example 6: Subspace of a Product Space

Consider the product space $X = \mathbb{R}^2$ with the product topology. Let $Y = \{(x, y) \in \mathbb{R}^2 \mid x = 0\}$, the y-axis. The subspace topology on Y is the topology inherited from $\mathbb{R}^2$, which is the standard topology on $\mathbb{R}$, since for every open set in $\mathbb{R}^2$, the intersection with the y-axis is open in $\mathbb{R}$.

2.2 Product Topology

2.2.1 Definition

Given two topological spaces (X, τ_X) and (Y, τ_Y), the **product topology** on the Cartesian product $X \times Y$ is the topology generated by the basis

$$\mathcal{B} = \{U \times V \mid U \in \tau_X, V \in \tau_Y\}.$$

A set $W \subseteq X \times Y$ is open in the product topology if and only if it can be written as a union of sets of the form $U \times V$, where U is open in X and V is open in Y.

The product topology is the topology that is most natural when we consider product spaces, which are spaces formed by taking the Cartesian product of two or more topological spaces. In this context, the product topology ensures that the projection maps from the product space to the individual spaces are continuous.

Examples

Example 1: Product of Two Real Lines

Consider $X = \mathbb{R}$ and $Y = \mathbb{R}$, both with the standard topology. The product topology on $\mathbb{R} \times \mathbb{R}$ is the standard topology on $\mathbb{R}^2$. The basic open sets in this topology are of the form $(a, b) \times (c, d)$, where (a, b) is an open interval in $\mathbb{R}$ and (c, d) is another open interval in $\mathbb{R}$. These basic open sets form a basis for the topology on $\mathbb{R}^2$. Thus, the product topology on $\mathbb{R} \times \mathbb{R}$ coincides with the standard topology on $\mathbb{R}^2$.

Example 2: Discrete and Standard Topology

Let $X = \{0, 1\}$ with the discrete topology and $Y = \mathbb{R}$ with the standard topology. In this case, the product topology on $X \times Y$ consists of sets of the form $\{0\} \times U \cup \{1\} \times V$, where U and V are open sets in $\mathbb{R}$. Since $X = \{0, 1\}$ has the discrete topology, every subset of X is open. Thus, the open sets in the product topology are unions of "rectangles" where one factor is a point in X and the other factor is an open set in $\mathbb{R}$.

Example 3: Product of a Circle and the Real Line

Let $X = \mathbb{S}^1$ (the unit circle) and $Y = \mathbb{R}$ with their standard topologies. The product space $\mathbb{S}^1 \times \mathbb{R}$ has a topology where the open sets are unions of sets of the form $U \times V$, with U open in $\mathbb{S}^1$ and V open in $\mathbb{R}$. For example, an open set in this product topology could be a set like $(\theta_1, \theta_2) \times (a, b)$, where (θ_1, θ_2) is an open interval on the unit circle and (a, b) is an open interval on the real line.

Example 4: Product of Two Compact Spaces

Consider $X = [0, 1]$ and $Y = [0, 1]$, each equipped with the standard topology. The product topology on $X \times Y = [0, 1] \times [0, 1]$ is the standard topology on the unit square in $\mathbb{R}^2$. The basic open sets are of the form $(a, b) \times (c, d)$, where (a, b) and (c, d) are open intervals in $[0, 1]$. Since $[0, 1]$ is compact and the product of compact spaces is compact, $[0, 1] \times [0, 1]$ is also compact in the product topology.

2.2.2 Basic Properties of Product Topology

- **Continuity of Projection Maps:** The projection maps $\pi_X : X \times Y \to X$ and $\pi_Y : X \times Y \to Y$ are continuous. The map π_X takes a point $(x, y) \in X \times Y$ and maps it to $x \in X$, and π_Y takes (x, y) to $y \in Y$. Since the product topology is generated by open sets of the form $U \times V$, and projections map such sets to open sets in X or Y, the projection maps are continuous.

- **Compactness:** If X and Y are compact, then $X \times Y$ is compact. This follows from the fact that the product of compact spaces is compact in the product topology. A general result of topology is that a product of compact spaces is compact under the product topology.

- **Hausdorff Property:** If X and Y are Hausdorff, then $X \times Y$ is Hausdorff. This is because in a Hausdorff space, distinct points can be separated by open sets. In the product topology, the product of two Hausdorff spaces is also Hausdorff, meaning distinct points in $X \times Y$ can be separated by open sets.

- **Coarseness of the Product Topology:** The product topology is the coarsest topology for which the projection maps π_X and π_Y are continuous. That is, it is the smallest topology on $X \times Y$ that makes the projections π_X and π_Y continuous.

Chapter 2: Subspace and Product Topology

2.2.3 Further Examples of Product Topologies

Example 5: Infinite Product of Real Lines

Consider the space $\mathbb{R}^{\mathbb{N}}$, the countable product of copies of $\mathbb{R}$, equipped with the product topology. The open sets in this topology are unions of sets of the form
$$U_1 \times U_2 \times U_3 \times \cdots$$
where $U_i \subset \mathbb{R}$ is open for each i, and only finitely many U_i are different from $\mathbb{R}$ (i.e., the sets are "almost" all $\mathbb{R}$, except for finitely many coordinates). This space is used in various areas, such as functional analysis and the study of product spaces in general.

Example 6: Product of Two Discrete Spaces

Let $X = \{a, b\}$ with the discrete topology and $Y = \{1, 2\}$ with the discrete topology. The product topology on $X \times Y$ will also be discrete, because every subset of $X \times Y$ is open. In the product topology, the open sets are all possible combinations of subsets of X and Y, which means the topology on $X \times Y$ is discrete.

Example 7: Product of a Compact and a Non-Compact Space

Let $X = [0, 1]$ (a compact space) and $Y = \mathbb{R}$ (a non-compact space). The product topology on $X \times Y$ is the standard topology on the rectangle $[0, 1] \times \mathbb{R}$. Even though Y is non-compact, the product space $X \times Y$ is not compact because $\mathbb{R}$ is non-compact. However, the subspace $[0, 1] \times [0, 1]$ would still be compact.

2.2.4 Tychonoff's Theorem

A central result in topology is **Tychonoff's Theorem**, which states that the arbitrary product of compact spaces is compact. This theorem is fundamental for various areas of topology, especially in the study of topological vector spaces and functional analysis.

2.3 Additional Basic Concepts

2.3.1 Closed Sets in Subspace Topology

A set $C \subset Y$ is closed in the subspace topology if and only if there exists a closed set $D \subset X$ such that $C = D \cap Y$. This follows from the definition of the subspace topology since closed sets in the subspace topology are precisely the intersections of closed sets in the parent space X with the subspace Y.

More formally, if $D \in \tau_X$ is closed in X, then $C = D \cap Y$ is closed in the subspace topology on Y.

2.3.2 Continuity in Subspace and Product Topology

A map $f : Z \to X$ is continuous if and only if $f^{-1}(U)$ is open in Z for every open set $U \subset X$. The definition of continuity remains the same in the context of subspace and product topologies, with specific modifications for the corresponding structures:

- **Continuity in Subspace Topology:** If $f : Z \to Y$ is a map into a subspace $Y \subseteq X$, then f is continuous with respect to the subspace topology on Y if and only if $f^{-1}(U \cap Y)$ is open in Z for every open set $U \subseteq X$. This reflects the fact that the open sets in the subspace topology on Y are the intersections of open sets from X with Y.

- **Continuity in Product Topology:** For a product space $X \times Y$, a map $f : Z \to X \times Y$ is continuous if and only if both projections $\pi_X \circ f$ and $\pi_Y \circ f$ are continuous. This condition ensures that the map behaves continuously with respect to the open sets in the product topology, which are generated by the open sets from each factor space.

2.3.3 Compactness and Hausdorff Spaces

The following are important properties related to compactness and Hausdorff spaces in the context of subspaces and product topologies:

- **Compactness in Subspaces:** If X is compact and $Y \subset X$ is closed in X, then Y is compact in the subspace topology. This is a standard result in topology, often referred to as the *subspace compactness theorem*. A closed subset of a compact space is compact in the subspace topology.

- **Tychonoff's Theorem:** A product of compact spaces is compact. This is one of the most important theorems in topology, often called Tychonoff's theorem. It states that the product of any collection of compact spaces, under the product topology, is compact. This theorem is foundational in areas such as functional analysis and general topology.

- **Hausdorff Property in Product Spaces:** If X and Y are Hausdorff spaces, then the product space $X \times Y$ is Hausdorff. This property is essential because it ensures that distinct points in the product space can be separated by disjoint open sets, making it a desirable topological property for many applications.

2.4 Further Concepts in Subspace and Product Topology

2.4.1 Open Sets in Subspace Topology

The subspace topology is defined by the open sets of the parent space X. Given a topological space (X, τ) and a subset $Y \subseteq X$, the open sets in the subspace topology on Y are the intersections of open sets from X with Y. Formally, a set $U \subseteq Y$ is open in the subspace topology if and only if there exists an open set $O \in \tau_X$ such that $U = O \cap Y$.

Example:
Consider $X = \mathbb{R}$ with the standard topology, and let $Y = (0, 1) \subset \mathbb{R}$. The open sets in the subspace topology on Y are the sets of the form $U \cap (0, 1)$ where U is an open set in $\mathbb{R}$. Thus, intervals such as $(\frac{1}{4}, \frac{3}{4})$ are open in the subspace topology.

2.4.2 Product of Subspaces

Let X_1 and X_2 be topological spaces, and let $Y_1 \subset X_1$ and $Y_2 \subset X_2$ be subspaces. The product space $Y_1 \times Y_2$ inherits a subspace topology from the product space $X_1 \times X_2$. In other words, the subspace topology on $Y_1 \times Y_2 \subset X_1 \times X_2$ is the topology generated by the intersections of open sets in $X_1 \times X_2$ with $Y_1 \times Y_2$.

Example:
Let $X_1 = \mathbb{R}$ with the standard topology, $X_2 = \mathbb{R}$ with the standard topology, $Y_1 = (0, 1)$, and $Y_2 = [1, 2]$. The subspace topology on $Y_1 \times Y_2 = (0, 1) \times [1, 2]$ is the topology inherited from $\mathbb{R}^2$, consisting of open sets of the form $O_1 \times O_2$, where $O_1 \subseteq \mathbb{R}$ is open in X_1 and $O_2 \subseteq \mathbb{R}$ is open in X_2, and $O_1 \times O_2$ intersects with $(0, 1) \times [1, 2]$.

2.4.3 Comparison of Subspace Topology and Product Topology

While both the subspace topology and the product topology allow us to generate new topological spaces from existing ones, they are distinct in how they structure open sets:

1. In the subspace topology, open sets are intersections of open sets from the parent space with the subspace.

2. In the product topology, open sets are generated by Cartesian products of open sets from each factor space.

These distinctions affect how continuous functions, compactness, and convergence behave in subspaces and product spaces.

Key Differences:
- The subspace topology reflects the topology of a subset of a space, with open sets intersecting the open sets of the parent space.

- The product topology reflects the topology of a product of spaces, where the open sets come from combinations of open sets in the individual spaces.

- In product spaces, open sets are more "structured" as products of open sets in each factor space.

2.4.4 Projections in Product Topology

In the product topology, the projection maps $\pi_X : X \times Y \to X$ and $\pi_Y : X \times Y \to Y$, which map a point $(x, y) \in X \times Y$ to $x \in X$ and $y \in Y$, are continuous by the definition of the product topology. These projections play a crucial role in many topological constructions and in understanding continuity in product spaces.

2.4.5 Subspace and Product Topology on Infinite Products

When working with infinite products, the subspace topology and the product topology behave similarly, but there are additional considerations:

- For an infinite product of spaces, say $\prod_{i \in I} X_i$, the product topology is generated by the basis of sets of the form $\prod_{i \in I} U_i$, where U_i is open in X_i for each i, and $U_i = X_i$ for all but finitely many i.

- The subspace topology on a subproduct space $\prod_{i \in I} Y_i$ (where each Y_i is a subspace of X_i) is the topology induced by the product topology on $\prod_{i \in I} X_i$.

Chapter 2: Subspace and Product Topology

Example:
Consider the product of two spaces, $X = \mathbb{R}$ and $Y = \mathbb{R}$. If $Z \subseteq X \times Y$ is a subspace, then the subspace topology on Z is the same as the topology inherited from $\mathbb{R}^2$. If X is the product of infinitely many spaces, say $\mathbb{R}^\mathbb{N}$, the product topology on this space will be the coarsest topology that makes all projections continuous, with each U_i being open in each copy of $\mathbb{R}$.

Summary of Key Properties of Subspace and Product Topology

- **Subspace Topology:** A set $A \subseteq Y$ is open in the subspace topology if and only if there exists an open set O in X such that $A = O \cap Y$.

- **Product Topology:** The product topology on $X \times Y$ is generated by open sets of the form $U \times V$, where $U \in \tau_X$ and $V \in \tau_Y$.

- **Continuity:** A function between product or subspace spaces is continuous if the preimage of every open set is open, with adjustments for the type of topology (subspace or product).

- **Compactness:** A subspace of a compact space is compact, and the product of compact spaces is compact.

- **Hausdorff:** The product of Hausdorff spaces is Hausdorff, and subspaces of Hausdorff spaces are Hausdorff.

2.5 MCQs: Subspace and Product Topology

1. Let X be a topological space and $A \subseteq X$. The subspace topology on A is defined by:

    (a) A inherits the topology of X directly
    (b) The collection of open sets in A is the intersection of open sets in X with A
    (c) A has the discrete topology
    (d) A has the indiscrete topology

    Answer: (b)
    Explanation: In the subspace topology, the open sets in A are the intersections of the open sets in X with the subset A.

2. Given two topological spaces X and Y, the product topology on $X \times Y$ is defined as:

   (a) The smallest topology containing all products of open sets from X and Y

   (b) The largest topology containing all products of open sets from X and Y

   (c) The discrete topology on $X \times Y$

   (d) The union of the topologies on X and Y

   Answer: (a)
   Explanation: The product topology on $X \times Y$ is the coarsest topology such that the projection maps are continuous, and open sets are of the form $U \times V$ where U is open in X and V is open in Y.

3. In the subspace topology, if $A \subseteq X$, and $U \subseteq A$ is open in the subspace topology, then:

   (a) U is necessarily open in X

   (b) U may not be open in X, but there exists some open set in X whose intersection with A is U

   (c) U is closed in X

   (d) U must be open and closed in X

   Answer: (b)
   Explanation: In the subspace topology, an open set in A is the intersection of A with an open set from the ambient space X.

4. Consider the product topology on $X \times Y$. A set $U \subseteq X \times Y$ is open in the product topology if and only if:

   (a) $U = U_1 \times U_2$ where U_1 is open in X and U_2 is open in Y

   (b) U is the union of sets of the form $U_1 \times U_2$, where U_1 is open in X and U_2 is open in Y

   (c) U is open in either X or Y

   (d) U is closed in $X \times Y$

Answer: (b)
Explanation: In the product topology, open sets are unions of basis elements, which are products of open sets from X and Y.

5. Let X be a topological space and $A \subseteq X$. The subspace topology on A is defined by:

    (a) A inherits the topology of X directly

    (b) The collection of open sets in A is the intersection of open sets in X with A

    (c) A has the discrete topology

    (d) A has the indiscrete topology

Answer: (b)
Explanation: In the subspace topology, the open sets in A are the intersections of the open sets in X with the subset A.

6. Let X be a topological space and $A \subseteq X$. If A is open in X, then in the subspace topology on A:

    (a) A is open in X

    (b) A is closed in X

    (c) A is open in the subspace topology

    (d) A must be both open and closed in X

Answer: (c)
Explanation: In the subspace topology, the open sets are intersections of open sets in X with A. If A is open in X, then A is open in the subspace topology.

7. The product topology on $X \times Y$ is defined by:

    (a) The finest topology making the projection maps continuous

    (b) The coarsest topology making the projection maps continuous

    (c) The topology where $X \times Y$ is a disjoint union of X and Y

    (d) The union of all topologies on X and Y

Answer: (b)
Explanation: The product topology is the coarsest topology on $X \times Y$ that makes the projection maps continuous.

8. Which of the following is an example of a basis element in the product topology on $X \times Y$?

    (a) A set of the form $U \times V$ where U is open in X and V is open in Y

    (b) A set of the form $U \cup V$ where U is open in X and V is open in Y

    (c) A set of the form $X \cap Y$

    (d) A set of the form $U \cap V$ where U is open in X and V is open in Y

Answer: (a)
Explanation: Basis elements of the product topology are Cartesian products of open sets from each factor space.

9. If A and B are subsets of X and Y respectively, which of the following sets is open in the product topology on $X \times Y$?

    (a) $A \times B$ where A is open in X and B is open in Y

    (b) $A \cup B$ where A is open in X and B is open in Y

    (c) $A \cap B$ where A is open in X and B is open in Y

    (d) $A \times B$ where A is closed in X and B is closed in Y

Answer: (a)
Explanation: A product $A \times B$ where A and B are open in their respective spaces is an open set in the product topology.

10. The product topology on $X \times Y$ is finer than:

    (a) The discrete topology on $X \times Y$

    (b) The indiscrete topology on $X \times Y$

    (c) The subspace topology induced by the discrete topology on X and Y

Chapter 2: Subspace and Product Topology

(d) The subspace topology induced by the indiscrete topology on X and Y

Answer: (b)
Explanation: The product topology is coarser (less fine) than the discrete topology and finer than the indiscrete topology.

11. Which of the following is true about the product of two topological spaces (X, τ_X) and (Y, τ_Y) under the product topology?

    (a) The product topology on $X \times Y$ has more open sets than the product of the topologies τ_X and τ_Y

    (b) The product topology on $X \times Y$ is the same as the topology on X and Y combined

    (c) The product topology on $X \times Y$ is the finest topology making all projection maps continuous

    (d) The product topology on $X \times Y$ is always discrete

Answer: (c)
Explanation: The product topology is the finest topology on $X \times Y$ that ensures the continuity of projection maps.

12. In the subspace topology, if A is closed in X, which of the following statements is true?

    (a) A is closed in the subspace topology

    (b) A is open in the subspace topology

    (c) A is both open and closed in the subspace topology

    (d) A is closed in X but not necessarily in the subspace topology

Answer: (d)
Explanation: If A is closed in X, it may not be closed in the subspace topology; it depends on how A intersects with other sets.

13. If X and Y are topological spaces and $A \subseteq X$ and $B \subseteq Y$, what is the subspace topology on $A \times B$ when viewed as a subset of $X \times Y$?

    (a) The subspace topology on $A \times B$ is generated by the basis sets $U \times V$ where U is open in A and V is open in B

(b) The subspace topology on $A \times B$ is generated by the basis sets $U \times V$ where U is open in X and V is open in Y

(c) The subspace topology on $A \times B$ is the discrete topology

(d) The subspace topology on $A \times B$ is the indiscrete topology

Answer: (b)
Explanation: The subspace topology on $A \times B$ is generated by the intersections of open sets in $X \times Y$ with $A \times B$.

14. The product of two compact topological spaces X and Y is:

    (a) Always compact

    (b) Always connected

    (c) Always Hausdorff

    (d) Always locally compact

Answer: (a)
Explanation: The product of two compact spaces is compact in the product topology.

15. In the product topology on $X \times Y$, if $U \subseteq X \times Y$ is open, then for each $(x, y) \in U$, there exists:

    (a) An open set $V \subseteq X$ and an open set $W \subseteq Y$ such that $(x, y) \in V \times W \subseteq U$

    (b) An open set $V \subseteq X$ such that $(x, y) \in V \times Y \subseteq U$

    (c) An open set $W \subseteq Y$ such that $(x, y) \in X \times W \subseteq U$

    (d) A closed set $V \subseteq X$ and a closed set $W \subseteq Y$ such that $(x, y) \in V \times W \subseteq U$

Answer: (a)
Explanation: For any open set U in the product topology, around every point $(x, y) \in U$, there exist open sets $V \subseteq X$ and $W \subseteq Y$ such that (x, y) is in the product $V \times W \subseteq U$.

16. In the product topology, if X and Y are both Hausdorff spaces, then $X \times Y$ is:

    (a) Hausdorff

Chapter 2: Subspace and Product Topology

(b) Not necessarily Hausdorff

(c) Compact

(d) Discrete

Answer: (a)

Explanation: The product of two Hausdorff spaces is Hausdorff in the product topology.

17. For the subspace topology on $A \subseteq X$, which of the following is true?

    (a) Every open set in A is the intersection of A with an open set in X

    (b) Every closed set in A is the intersection of A with a closed set in X

    (c) Every open set in X is an open set in A

    (d) Every closed set in X is a closed set in A

Answer: (a)

Explanation: In the subspace topology, open sets in A are precisely the intersections of A with open sets in X.

18. The product of two connected spaces X and Y is:

    (a) Connected

    (b) Disconnected

    (c) Compact

    (d) Discrete

Answer: (a)

Explanation: The product of two connected spaces is connected.

19. The subspace topology on $A \subseteq X$ can be:

    (a) Finer than the topology on X

    (b) Coarser than the topology on X

    (c) The same as the topology on X

    (d) Always discrete

Answer: (b)
Explanation: The subspace topology is always coarser than or equal to the topology on the ambient space X.

20. For the product topology on $X \times Y$, which of the following is a basis for the topology?

    (a) Sets of the form $U \times V$, where U is open in X and V is open in Y
    (b) Sets of the form $U \cup V$, where U is open in X and V is open in Y
    (c) Sets of the form $X \cap Y$
    (d) Sets of the form $U \cap V$, where U is open in X and V is open in Y

Answer: (a)
Explanation: Basis elements for the product topology are Cartesian products of open sets from each space.

21. If X and Y are topological spaces and $A \subseteq X$ and $B \subseteq Y$, then $A \times B$ is:

    (a) Open in $X \times Y$ if A and B are open in X and Y, respectively
    (b) Closed in $X \times Y$ if A and B are closed in X and Y, respectively
    (c) Compact if A and B are compact
    (d) Connected if A and B are connected

Answer: (a)
Explanation: $A \times B$ is open in $X \times Y$ if A and B are open in X and Y, respectively.

22. In the subspace topology, if $A \subseteq X$ is an open set in X, then:

    (a) A is open in X and therefore in the subspace topology
    (b) A is closed in the subspace topology
    (c) A is always both open and closed in X
    (d) A is not necessarily open in the subspace topology

Answer: (a)
Explanation: If A is open in X, then it is open in the subspace topology.

Chapter 2: Subspace and Product Topology

23. The product topology on $X \times Y$ is such that:

    (a) The projection maps are continuous
    (b) The projection maps are not necessarily continuous
    (c) It is the same as the topology generated by all open sets of X and Y
    (d) The topology is always discrete

    Answer: (a)
    Explanation: The product topology is designed so that the projection maps are continuous.

24. If $X \times Y$ is compact, then:

    (a) X and Y are both compact
    (b) X and Y are not necessarily compact
    (c) X is compact, but Y is not necessarily compact
    (d) Y is compact, but X is not necessarily compact

    Answer: (a)
    Explanation: The compactness of $X \times Y$ implies that both X and Y must be compact.

25. In the subspace topology on $A \subseteq X$, if A is a closed set in X, then:

    (a) A is closed in the subspace topology
    (b) A is open in the subspace topology
    (c) A is necessarily open and closed in X
    (d) A may not be closed in the subspace topology

    Answer: (a)
    Explanation: If A is closed in X, then it is closed in the subspace topology on A.

26. If A and B are subsets of X and Y respectively, then:

    (a) The intersection $A \cap B$ is open in the product topology
    (b) The union $A \cup B$ is open in the product topology

(c) The product $A \times B$ is open in the product topology if A and B are open

(d) The product $A \times B$ is closed in the product topology if A and B are closed

Answer: (c)
Explanation: $A \times B$ is open in the product topology if A and B are open in their respective spaces.

27. In the product topology, the space $X \times Y$ is:

    (a) Compact if X and Y are compact

    (b) Connected if X and Y are connected

    (c) Discrete if X and Y are discrete

    (d) Hausdorff if X and Y are Hausdorff

    Answer: all are correct

28. If X and Y are topological spaces and $X \times Y$ is connected, then:

    (a) Both X and Y must be connected

    (b) Both X and Y need not be connected

    (c) Both X and Y must be compact

    (d) Both X and Y must be Hausdorff

    Answer: (a)
    Explanation: The connectedness of $X \times Y$ does necessarily imply the connectedness of X and Y individually.

29. For the subspace topology, the open sets in $A \subseteq X$ are:

    (a) Intersections of open sets in X with A

    (b) Unions of closed sets in X with A

    (c) All sets of the form $U \cup A$ where U is open in X

    (d) All sets of the form $X \cap A$ where X is open in X

Chapter 2: Subspace and Product Topology

Answer: (a)
Explanation: The open sets in the subspace topology are intersections of open sets in the ambient space with the subset.

30. The product topology on $X \times Y$ is:

    (a) The coarsest topology making the projection maps continuous

    (b) The finest topology making the projection maps continuous

    (c) The same as the topology on X and Y

    (d) Always compact

Answer: (b)
Explanation: The product topology is the finest topology that makes the projection maps continuous.

31. If $X \times Y$ is Hausdorff, then:

    (a) X and Y are Hausdorff

    (b) X and Y need not be Hausdorff

    (c) X is Hausdorff, but Y is not necessarily Hausdorff

    (d) Y is Hausdorff, but X is not necessarily Hausdorff

Answer: (a)
Explanation: The product of Hausdorff spaces is Hausdorff.

32. In the subspace topology, if $A \subseteq X$ is compact, then:

    (a) A is compact in the subspace topology

    (b) A is not necessarily compact in the subspace topology

    (c) A is always open in the subspace topology

    (d) A is always closed in the subspace topology

Answer: (a)
Explanation: Compactness is preserved in the subspace topology.

33. For the product topology on $X \times Y$, the basis elements are:

    (a) Products of open sets from X and Y

(b) Unions of open sets from X and Y

(c) Intersections of open sets from X and Y

(d) Products of closed sets from X and Y

Answer: (a)
Explanation: Basis elements in the product topology are Cartesian products of open sets from each space.

34. The subspace topology on $A \subseteq X$ is such that:

    (a) Every open set in A is open in X

    (b) Every open set in A is an intersection of A with an open set in X

    (c) Every open set in X is open in A

    (d) Every closed set in X is closed in A

    Answer: (b)
    Explanation: Open sets in A are intersections of A with open sets in X.

35. The product topology on $X \times Y$ is:

    (a) The same as the topology generated by all possible open sets in $X \times Y$

    (b) Finer than the topology generated by Cartesian products of open sets from X and Y

    (c) Coarser than the topology generated by Cartesian products of open sets from X and Y

    (d) The discrete topology on $X \times Y$

    Answer: (a)
    Explanation: The product topology is the same as the topology generated by all possible open sets in $X \times Y$.

36. If $A \times B$ is open in $X \times Y$, then:

    (a) A and B are both open in X and Y

    (b) A and B are both closed in X and Y

    (c) A and B are compact in X and Y

Chapter 2: Subspace and Product Topology 85

   (d) A and B are connected in X and Y

   Answer: (a)
   Explanation: In the product topology, a product $A \times B$ is open if A is open in X and B is open in Y.

37. If X and Y are both Hausdorff, then $X \times Y$ is:

    (a) Not necessarily compact

    (b) Always connected

    (c) Always Hausdorff

    (d) Always discrete

    Answer: (c)
    Explanation: The product of Hausdorff spaces is Hausdorff.

38. The product topology on $X \times Y$ ensures that:

    (a) All projection maps are continuous

    (b) All projection maps are not necessarily continuous

    (c) The topology is always discrete

    (d) The topology is always coarser than any topology on $X \times Y$

    Answer: (a)
    Explanation: The product topology is defined to make the projection maps continuous.

39. The subspace topology on $A \subseteq X$ where A is open in X is:

    (a) The same as the subspace topology on A

    (b) Always discrete

    (c) Always Hausdorff

    (d) Always compact

    Answer: (a)
    Explanation: If A is open in X, the subspace topology on A is the intersection of A with open sets in X.

40. For two topological spaces X and Y, the product topology on $X \times Y$ has:

    (a) Fewer open sets than the product of the topologies on X and Y

    (b) More open sets than the product of the topologies on X and Y

    (c) The same number of open sets as the product of the topologies on X and Y

    (d) The same number of open sets as the topology on X and Y

    Answer: (c)
    Explanation: The product topology has same number of open sets as the product of the topologies on X and Y.

41. If $X \times Y$ is compact, then:

    (a) X and Y must be compact

    (b) X and Y need not be compact

    (c) X is compact but Y is not necessarily compact

    (d) Y is compact but X is not necessarily compact

    Answer: (a)
    Explanation: The compactness of $X \times Y$ implies the compactness of both X and Y.

42. In the subspace topology, if $A \subseteq X$ and A is compact in X, then:

    (a) A is compact in the subspace topology

    (b) A is not necessarily compact in the subspace topology

    (c) A is always open in the subspace topology

    (d) A is always closed in the subspace topology

    Answer: (a)
    Explanation: Compactness is preserved when moving to the subspace topology.

43. The subspace topology on $A \subseteq X$ has open sets that are:

    (a) Intersections of A with open sets from X

Chapter 2: Subspace and Product Topology

(b) Intersections of A with closed sets from X

(c) Unions of A with open sets from X

(d) Unions of A with closed sets from X

Answer: (a)
Explanation: Open sets in the subspace topology are intersections of A with open sets in X.

44. The product topology on $X \times Y$ is:

    (a) The same as the topology generated by Cartesian products of open sets from X and Y

    (b) Coarser than the topology generated by Cartesian products of open sets from X and Y

    (c) Finer than the topology generated by Cartesian products of open sets from X and Y

    (d) Always discrete

Answer: (a)
Explanation: The product topology is the same as the topology generated by Cartesian products of open sets from X and Y.

45. In the product topology, if X is compact and Y is not compact, then:

    (a) $X \times Y$ is not compact

    (b) $X \times Y$ is compact

    (c) $X \times Y$ is connected

    (d) $X \times Y$ is Hausdorff

Answer: (a)
Explanation: The product of a compact space with a non-compact space is not compact.

46. If $A \subseteq X$ and $B \subseteq Y$ are closed sets, then:

    (a) $A \times B$ is closed in $X \times Y$

    (b) $A \times B$ is open in $X \times Y$

    (c) $A \times B$ is not necessarily closed in $X \times Y$

(d) $A \times B$ is connected

Answer: (a)
Explanation: The product of closed sets is closed in the product topology.

47. In the subspace topology, if $A \subseteq X$ and A is open in X, then:

    (a) A is open in the subspace topology
    (b) A is closed in the subspace topology
    (c) A is neither open nor closed in the subspace topology
    (d) A is open and closed in the subspace topology

Answer: (a)
Explanation: If A is open in X, then A is open in the subspace topology.

48. For the product topology on $X \times Y$, which of the following is true?

    (a) The product of two open sets is not necessarily open
    (b) The product of two open sets is always open
    (c) The product of two closed sets is always closed
    (d) The product of two closed sets is not necessarily closed

Answer: (b)
Explanation: The product of two open sets is open, but the product of two closed sets is not necessarily closed.

49. In the product topology, if $X \times Y$ is connected, then:

    (a) Both X and Y must be connected
    (b) At least one of X or Y must be connected
    (c) Neither X nor Y needs to be connected
    (d) Both X and Y must be compact

Answer: (a)
Explanation: If $X \times Y$ is connected in product topology then both X and Y must be connected.

Chapter 2: Subspace and Product Topology

50. If X and Y are topological spaces, the product topology on $X \times Y$ is:

    (a) Finer than the product of the topologies on X and Y
    (b) Coarser than the product of the topologies on X and Y
    (c) The same as the product of the topologies on X and Y
    (d) Discrete

    Answer: (c)
    Explanation: The product topology is the same as the product of the topologies on X and Y.

51. The product topology is:

    (a) The finest topology making projection maps continuous
    (b) The coarsest topology making projection maps continuous
    (c) The discrete topology
    (d) The same as the topology generated by open sets in X and Y

    Answer: (b)
    Explanation: The product topology is the coarsest topology that makes the projection maps continuous.

52. In the product topology on $X \times Y$, if U is open in X and V is open in Y, then:

    (a) $U \times V$ is open in $X \times Y$
    (b) $U \times V$ is not necessarily open in $X \times Y$
    (c) $U \times V$ is closed in $X \times Y$
    (d) $U \times V$ is connected in $X \times Y$

    Answer: (a)
    Explanation: The product of open sets is open in the product topology.

53. For the product topology on $X \times Y$, if X is discrete and Y is compact, then:

    (a) $X \times Y$ is compact
    (b) $X \times Y$ is not compact

(c) $X \times Y$ is connected

(d) $X \times Y$ is Hausdorff

Answer: (b)
Explanation: The product of a discrete space with a compact space is not necessarily compact.

54. If X is a Hausdorff space and Y is compact, then $X \times Y$ is:

    (a) Hausdorff

    (b) Not Hausdorff

    (c) Compact

    (d) Connected

Answer: (a)
Explanation: The product of a Hausdorff space with a compact space is Hausdorff.

55. If $X \times Y$ is Hausdorff, then:

    (a) Both X and Y are Hausdorff

    (b) Only X is Hausdorff

    (c) Only Y is Hausdorff

    (d) Neither X nor Y is necessarily Hausdorff

Answer: (a)
Explanation: The product of Hausdorff spaces is Hausdorff.

56. For the subspace topology on $A \subseteq X$, if A is closed in X, then:

    (a) A is closed in the subspace topology

    (b) A is open in the subspace topology

    (c) A is neither open nor closed in the subspace topology

    (d) A is always open and closed in the subspace topology

Answer: (d)
Explanation: In any topological space, the whole space and the empty set are always both open and closed.

Chapter 2: Subspace and Product Topology

57. In the product topology, if X is connected and Y is not connected, then:

    (a) $X \times Y$ may or may not be connected

    (b) $X \times Y$ is connected

    (c) $X \times Y$ is not connected

    (d) $X \times Y$ is compact

    Answer: (c)
    Explanation: The connectedness of $X \times Y$ depends on both X and Y.

58. For the product topology, the projection maps:

    (a) Are continuous

    (b) Are not necessarily continuous

    (c) Are always discrete

    (d) Are not related to the product topology

    Answer: (a)
    Explanation: The product topology is defined to make the projection maps continuous.

59. If X and Y are compact Hausdorff spaces, then $X \times Y$ is:

    (a) Compact and Hausdorff

    (b) Not necessarily compact

    (c) Not necessarily Hausdorff

    (d) Always connected

    Answer: (a)
    Explanation: The product of compact Hausdorff spaces is compact and Hausdorff.

60. In the subspace topology on $A \subseteq X$, if A is neither open nor closed in X, then:

    (a) A may be open or closed in the subspace topology

(b) A is necessarily open in the subspace topology

(c) A is necessarily closed in the subspace topology

(d) A is neither open nor closed in the subspace topology

Answer: (a)
Explanation: The properties of openness and closedness in the subspace topology can differ from those in the ambient space.

61. The subspace topology on a subset $Y \subseteq X$ of a topological space X has open sets:

    (a) $\{U \cap Y \mid U \text{ is open in } X\}$

    (b) All open sets in X

    (c) All subsets of Y

    (d) $\{U \subseteq Y \mid U \text{ is open in } X\}$

Answer: (a)
Explanation: The subspace topology on Y consists of intersections of open sets in X with Y.

62. A set $A \subseteq Y$ is open in the subspace topology on $Y \subseteq X$ if:

    (a) There exists an open set $U \subseteq X$ such that $A = U \cap Y$

    (b) A is open in X

    (c) A is closed in X

    (d) $A = Y$

Answer: (a)
Explanation: By definition, open sets in Y are of the form $U \cap Y$, where U is open in X.

63. A set $A \subseteq Y$ is closed in the subspace topology on $Y \subseteq X$ if:

    (a) There exists a closed set $C \subseteq X$ such that $A = C \cap Y$

    (b) A is closed in X

    (c) A is open in Y

    (d) $A = \emptyset$

Chapter 2: Subspace and Product Topology

Answer: (a)
Explanation: A set is closed in Y if it is the intersection of a closed set in X with Y.

64. If $Y \subseteq X$ is a subspace and X is Hausdorff, then Y is:

    (a) Hausdorff
    (b) Not necessarily Hausdorff
    (c) Compact
    (d) Connected

    Answer: (a)
    Explanation: The subspace topology inherits the Hausdorff property, as disjoint open sets in X intersect Y to separate points.

65. If $Y \subseteq X$ is a subspace and X is compact, then Y is compact if:

    (a) Y is closed in X
    (b) Y is open in X
    (c) Y is dense in X
    (d) $Y = X$

    Answer: (a)
    Explanation: A closed subset of a compact space is compact.

66. If $Y \subseteq X$ is a subspace and X is connected, then Y is:

    (a) Not necessarily connected
    (b) Connected
    (c) Compact
    (d) Hausdorff

    Answer: (a)
    Explanation: Connectedness is not always inherited; e.g., $Y = (0,1) \cup (2,3) \subset \mathbb{R}$ is not connected.

67. A function $f : Y \to Z$, where $Y \subseteq X$ is a subspace, is continuous if:

(a) $f : Y \to Z$ is continuous with respect to the subspace topology on Y

(b) f is continuous as a map from $X \to Z$

(c) f is bijective

(d) f is surjective

Answer: (a)

Explanation: Continuity is defined with respect to the subspace topology on Y.

68. The inclusion map $i : Y \to X$, where $Y \subseteq X$ is a subspace, is:

    (a) Continuous

    (b) Not continuous

    (c) Bijective

    (d) Open

Answer: (a)

Explanation: The inclusion map is continuous, as preimages of open sets in X are open in Y.

69. If $A \subseteq Y \subseteq X$, the closure of A in Y is:

    (a) $\text{Cl}_Y(A) = \text{Cl}_X(A) \cap Y$

    (b) $\text{Cl}_X(A)$

    (c) A

    (d) Y

Answer: (a)

Explanation: The closure in Y is the intersection of the closure in X with Y.

70. A subspace $Y \subseteq X$ is dense in X if:

    (a) $\text{Cl}_X(Y) = X$

    (b) Y is open

    (c) Y is compact

Chapter 2: Subspace and Product Topology

(d) Y is finite

Answer: (a)
Explanation: A set is dense if its closure is the entire space.

71. The product topology on $X \times Y$ has a basis consisting of:

    (a) Sets of the form $U \times V$, where U is open in X and V is open in Y
    (b) Sets of the form $U \times Y$
    (c) Sets of the form $X \times V$
    (d) All subsets of $X \times Y$

    Answer: (a)
    Explanation: The product topology is generated by products of open sets from X and Y.

72. An open set in the product topology on $X \times Y$ is:

    (a) A union of sets of the form $U \times V$, where U is open in X and V is open in Y
    (b) A single set $U \times V$
    (c) A closed set in $X \times Y$
    (d) The entire space $X \times Y$

    Answer: (a)
    Explanation: Open sets are arbitrary unions of basis elements $U \times V$.

73. The projection map $\pi_X : X \times Y \to X$, $\pi_X(x, y) = x$, is:

    (a) Continuous
    (b) Not continuous
    (c) Bijective
    (d) A homeomorphism

    Answer: (a)
    Explanation: Projections are continuous, as preimages of open sets are of the form $U \times Y$, which are open.

74. If X and Y are Hausdorff, then $X \times Y$ with the product topology is:

(a) Hausdorff

(b) Not necessarily Hausdorff

(c) Compact

(d) Connected

Answer: (a)
Explanation: The product of Hausdorff spaces is Hausdorff, as points can be separated by products of open sets.

75. Tychonoff's theorem states that:

    (a) The product of any collection of compact spaces is compact

    (b) The product of connected spaces is connected

    (c) The product of Hausdorff spaces is Hausdorff

    (d) The product of separable spaces is separable

    Answer: (a)
    Explanation: Tychonoff's theorem ensures that arbitrary products of compact spaces are compact in the product topology.

76. If X and Y are connected, then $X \times Y$ is:

    (a) Connected

    (b) Not necessarily connected

    (c) Compact

    (d) Hausdorff

    Answer: (a)
    Explanation: The product of connected spaces is connected, as $X \times Y$ cannot be split into disjoint open sets.

77. If X and Y are compact, then $X \times Y$ is:

    (a) Compact

    (b) Not necessarily compact

    (c) Connected

    (d) Separable

Chapter 2: Subspace and Product Topology

Answer: (a)
Explanation: By Tychonoff's theorem, the product of compact spaces is compact.

78. A subspace $Z \subseteq X \times Y$ is open if:

    (a) It is a union of sets of the form $(U \times V) \cap Z$, where $U \subset X$, $V \subset Y$ are open
    (b) It is closed in $X \times Y$
    (c) It is compact
    (d) It is the entire space

Answer: (a)
Explanation: Open sets in the subspace topology are intersections of open sets in $X \times Y$ with Z.

79. The product topology on $\prod_{i \in I} X_i$ has a basis consisting of:

    (a) Sets of the form $\prod_{i \in I} U_i$, where U_i is open in X_i for finitely many i
    (b) All sets $\prod_{i \in I} U_i$
    (c) Sets where all $U_i = X_i$
    (d) Finite products only

Answer: (a)
Explanation: The basis consists of sets where all but finitely many $U_i = X_i$.

80. If X is separable, then $X \times X$ is:

    (a) Separable
    (b) Not necessarily separable
    (c) Compact
    (d) Connected

Answer: (a)
Explanation: If X has a countable dense subset D, then $D \times D$ is countable and dense in $X \times X$.

81. A subspace $Y \subseteq X$ is compact if:

    (a) It is closed and X is compact
    (b) It is open
    (c) It is dense
    (d) It is infinite

    Answer: (a)
    Explanation: A closed subset of a compact space is compact.

82. The subspace $(0, 1) \subset \mathbb{R}$ with the standard topology is:

    (a) Not compact
    (b) Compact
    (c) Closed
    (d) Finite

    Answer: (a)
    Explanation: $(0, 1)$ is not compact, as it is not closed in $\mathbb{R}$.

83. The product $[0, 1] \times [0, 1] \subset \mathbb{R}^2$ is:

    (a) Compact
    (b) Not compact
    (c) Open
    (d) Infinite

    Answer: (a)
    Explanation: $[0, 1]$ is compact, so $[0, 1] \times [0, 1]$ is compact by Tychonoff's theorem.

84. A subspace $Y \subseteq X$ is connected if:

    (a) It cannot be written as the union of two disjoint non-empty open sets in Y
    (b) It is compact
    (c) It is closed

Chapter 2: Subspace and Product Topology

(d) It is open

Answer: (a)
Explanation: Connectedness in Y is defined with respect to the subspace topology.

85. The subspace $\mathbb{Q} \subset \mathbb{R}$ with the standard topology is:

    (a) Not connected
    (b) Connected
    (c) Compact
    (d) Closed

Answer: (a)
Explanation: $\mathbb{Q}$ is totally disconnected, as it can be split by irrational numbers.

86. The projection map $\pi_Y : X \times Y \to Y$ is:

    (a) Open
    (b) Closed
    (c) Bijective
    (d) A homeomorphism

Answer: (a)
Explanation: Projections are open, as the image of $U \times V$ is V, which is open.

87. If $A \subseteq X$ and $B \subseteq Y$, then the closure of $A \times B$ in $X \times Y$ is:

    (a) $\text{Cl}_X(A) \times \text{Cl}_Y(B)$
    (b) $A \times B$
    (c) $X \times Y$
    (d) $\text{Cl}_X(A) \times Y$

Answer: (a)
Explanation: The closure in the product topology is the product of the closures.

88. A subspace $Y \subseteq X$ is locally compact if:

    (a) Every point in Y has a compact neighborhood in Y

    (b) Y is compact

    (c) Y is connected

    (d) Y is open

    Answer: (a)
    Explanation: Local compactness is defined in the subspace topology.

89. The product $\mathbb{R} \times \mathbb{R}$ with the product topology is:

    (a) Locally compact

    (b) Not locally compact

    (c) Compact

    (d) Finite

    Answer: (a)
    Explanation: $\mathbb{R}$ is locally compact, and products of locally compact spaces are locally compact.

90. If X is second-countable, then $X \times X$ is:

    (a) Second-countable

    (b) Not second-countable

    (c) Compact

    (d) Connected

    Answer: (a)
    Explanation: The product of countable bases forms a countable basis for $X \times X$.

91. A subspace $Y \subseteq X$ is path-connected if:

    (a) Any two points in Y can be joined by a continuous path in Y

    (b) Y is connected

    (c) Y is compact

(d) Y is open

Answer: (a)
Explanation: Path-connectedness is defined in the subspace topology.

92. The subspace $S^1 \subset \mathbb{R}^2$ (the unit circle) is:

    (a) Path-connected

    (b) Not path-connected

    (c) Not connected

    (d) Open

Answer: (a)
Explanation: The circle is path-connected, as any two points can be joined by an arc.

93. The product $S^1 \times S^1$ (the torus) is:

    (a) Path-connected

    (b) Not path-connected

    (c) Not connected

    (d) Open

Answer: (a)
Explanation: The product of path-connected spaces is path-connected.

94. A subspace $Y \subseteq X$ is separable if:

    (a) It has a countable dense subset

    (b) It is compact

    (c) It is connected

    (d) It is open

Answer: (a)
Explanation: Separability is defined by a countable dense subset in Y.

95. If X is Lindelöf, then $X \times X$ is:

(a) Lindelöf

(b) Not necessarily Lindelöf

(c) Compact

(d) Connected

Answer: (a)
Explanation: Products of Lindelöf spaces are Lindelöf, as countable subcovers can be constructed.

96. The subspace topology on $Y \subseteq X$ ensures that a set $A \subseteq Y$ is compact if:

    (a) It is compact in X

    (b) It is open in Y

    (c) It is dense in Y

    (d) It is infinite

    Answer: (a)
    Explanation: Compactness in Y is equivalent to compactness in X.

97. The product topology on $\mathbb{R}^\omega$ (countable product of $\mathbb{R}$) is:

    (a) Not compact

    (b) Compact

    (c) Closed

    (d) Finite

    Answer: (a)
    Explanation: $\mathbb{R}^\omega$ is not compact, as $\mathbb{R}$ is not compact.

98. If $A \subseteq X$ is open, then $A \times Y$ is:

    (a) Open in $X \times Y$

    (b) Closed in $X \times Y$

    (c) Compact

    (d) Dense

Chapter 2: Subspace and Product Topology

Answer: (a)
Explanation: $A \times Y$ is a basic open set in the product topology.

99. The subspace $[0, 1) \subset \mathbb{R}$ is:

    (a) Not closed
    (b) Closed
    (c) Compact
    (d) Finite

Answer: (a)
Explanation: $[0, 1)$ is not closed, as 1 is a limit point not in the set.

100. The product $\prod_{n=1}^{\infty} [0, 1]$ is:

    (a) Compact
    (b) Not compact
    (c) Open
    (d) Connected

Answer: (a)
Explanation: By Tychonoff's theorem, the countable product of compact spaces is compact.

101. A subspace $Y \subseteq X$ is normal if:

    (a) X is normal and Y is closed
    (b) Y is open
    (c) Y is compact
    (d) Y is dense

Answer: (a)
Explanation: A closed subset of a normal space is normal.

102. The product $\mathbb{Q} \times \mathbb{Q} \subset \mathbb{R} \times \mathbb{R}$ is:

    (a) Not connected
    (b) Connected

(c) Compact

(d) Closed

Answer: (a)
Explanation: $\mathbb{Q}$ is totally disconnected, so $\mathbb{Q} \times \mathbb{Q}$ is also totally disconnected.

103. The subspace topology on $Y \subseteq X$ ensures that a function $f : Z \to Y$ is continuous if:

    (a) The composition $i \circ f : Z \to X$ is continuous, where $i : Y \to X$ is the inclusion

    (b) f is continuous in X

    (c) f is bijective

    (d) f is surjective

Answer: (a)
Explanation: Continuity is defined with respect to the subspace topology on Y.

104. The product topology on $X_1 \times X_2 \times \cdots \times X_n$ has open sets generated by:

    (a) Products $U_1 \times U_2 \times \cdots \times U_n$, where each U_i is open in X_i

    (b) Single sets $U_i \times X_j$

    (c) Closed sets

    (d) The entire space

Answer: (a)
Explanation: The basis consists of products of open sets.

Chapter 3

Separation Axioms

3.1 T_0 Spaces (Kolmogorov Spaces)

A topological space X is called a T_0 **space** or **Kolmogorov space** if for every pair of distinct points $x, y \in X$, there is at least one open set that contains one of the points and not the other. This is the weakest separation property.

Definition

A topological space X is said to be T_0 if for every pair of distinct points $x \neq y$, there exists an open set $U \subseteq X$ such that either:

$$x \in U \text{ and } y \notin U \quad \text{or} \quad y \in U \text{ and } x \notin U.$$

In other words, there exists an open set that "distinguishes" between any two distinct points in the space.

Examples of T_0 Spaces

1. **Finite Discrete Space**: Let $X = \{a, b\}$ with the discrete topology $\tau = \{\emptyset, \{a\}, \{b\}, \{a, b\}\}$. This is clearly a T_0 space since we can separate a and b by the open sets $\{a\}$ and $\{b\}$. More generally, any discrete topology on a set is T_0, as each point can be distinguished by the open set containing just that point.

2. **Co-finite Topology**: Let $X = \mathbb{R}$ and let the open sets be the co-finite topology $\tau = \{\emptyset\} \cup \{U \subseteq X \mid X \setminus U \text{ is finite}\}$. This is a T_0 space since for any distinct $x, y \in \mathbb{R}$, there exists an open set that contains one but not the other. For example, we can take the open set $U = \mathbb{R} \setminus \{y\}$ to separate x and y.

3. **Sierpinski Space**: Let $X = \{a, b\}$ with $\tau = \{\emptyset, \{a\}, X\}$. This space is T_0 because a is in $\{a\}$ but b is not. Additionally, b is in $X = \{a, b\}$, which contains both a and b, making it another open set containing b but not a.

4. **The Rational Numbers $\mathbb{Q}$ with the Subspace Topology**: Let $X = \mathbb{R}$ and let $Y = \mathbb{Q}$ (the set of rational numbers). With the subspace topology inherited from $\mathbb{R}$, Y is a T_0 space. For any two distinct rational numbers $x, y \in \mathbb{Q}$, there exists an open set in $\mathbb{R}$ that separates them, and this open set, when intersected with $\mathbb{Q}$, remains open in the subspace topology on $\mathbb{Q}$.

5. **Indiscrete Topology on a Nontrivial Set**: Consider a set $X = \{a, b\}$ with the indiscrete topology, i.e., $\tau = \{\emptyset, X\}$. This space is trivially T_0 because the only open sets are X and $\emptyset$, and the open sets distinguish points based on the fact that they either contain both points or none.

Properties of T_0 Spaces

- T_0 spaces are closed under taking subspaces, i.e., any subspace of a T_0 space is also T_0.

- Every T_1 space is also T_0, and every T_2 (Hausdorff) space is also T_0.

- A T_0 space does not necessarily imply any form of separation in terms of closed sets or other stronger properties such as T_1 or T_2.

- A space with a discrete topology is always a T_0 space, as each point is separated by an open set containing only that point.

- If a space is T_0, it does not necessarily have to be connected or Hausdorff. For example, the cofinite topology on $\mathbb{R}$ is T_0, but it is not Hausdorff.

Relation with Other Separation Axioms

1. T_0 is the weakest of the separation properties in topology. It guarantees that distinct points are distinguishable by open sets, but it doesn't require that points can be separated by disjoint open sets.

2. The next level of separation is T_1, which requires that for any pair of distinct points, each point has a neighborhood not containing the other.

Chapter 3: Separation Axioms

3. T_2 (Hausdorff) requires that for any two distinct points, there exist disjoint neighborhoods around each point, providing a stronger form of separation.

Applications of T_0 Spaces

T_0 spaces arise frequently in theoretical aspects of topology, particularly in general topology and categorical topology. The concept of distinguishability between points, even without the full separation of points as in T_1 or T_2, can still be useful in certain mathematical contexts such as in the study of convergence of nets and in the construction of certain types of spaces in category theory.

Conclusion

In summary, a T_0 space is a very general type of space where distinct points can be distinguished by some open set. Although it does not provide the separation of points as in T_1 or T_2 spaces, the T_0 property forms the foundation for stronger separation axioms, and it remains an important concept in the study of topological properties and structures.

3.2 T_1 Spaces (Frechet Spaces)

A topological space X is called a T_1 **space** or **Frechet space** if for every pair of distinct points $x, y \in X$, each has a neighborhood that does not contain the other. In other words, singletons are closed sets.

Definition

A topological space X is said to be T_1 if for every pair of distinct points $x \neq y$, there exist open sets U and V such that:

$$x \in U \quad \text{and} \quad y \notin U, \quad \text{and} \quad y \in V \quad \text{and} \quad x \notin V.$$

This condition ensures that for any two distinct points, there are open sets separating them. A crucial consequence of this property is that in a T_1 space, singletons are closed sets. That is, for every point $x \in X$, the set $\{x\}$ is closed in X.

Examples of T_1 Spaces

1. **Discrete Topology**: Any space with the discrete topology is T_1 because every singleton $\{x\}$ is a closed set. In fact, in a discrete space, every subset is open, so singletons are trivially closed.

2. **Co-finite Topology on $\mathbb{R}$**: Consider the co-finite topology on $\mathbb{R}$, where the open sets are either $\emptyset$ or any subset of $\mathbb{R}$ whose complement is finite. In this topology, every singleton $\{x\}$ is closed because its complement is co-finite, which is an open set. Hence, the space is T_1.

3. **Finite Sets with Discrete Topology**: Let $X = \{a, b, c\}$ with the discrete topology. Every singleton $\{a\}$, $\{b\}$, and $\{c\}$ is closed in the discrete topology, making the space T_1. This is true for any finite set with the discrete topology.

4. **The Real Line $\mathbb{R}$ with the Standard Topology**: The standard topology on $\mathbb{R}$ is T_1 because for any distinct points $x, y \in \mathbb{R}$, we can find open sets around x and y that do not contain the other point. For instance, $U = (x - \epsilon, x + \epsilon)$ and $V = (y - \epsilon, y + \epsilon)$ for some small $\epsilon > 0$ will separate x and y, implying that singletons are closed.

5. **The Sierpinski Space**: Let $X = \{a, b\}$ with $\tau = \{\emptyset, \{a\}, X\}$. This is a T_1 space because the singleton $\{a\}$ is closed, and the singleton $\{b\}$ is also closed.

6. **Any Finite Topological Space with the Discrete Topology**: If $X = \{x_1, x_2, \ldots, x_n\}$ with the discrete topology, every singleton set $\{x_i\}$ is closed, and hence, the space is T_1.

Properties of T_1 Spaces

- In a T_1 space, singletons are closed. This is one of the key properties that distinguishes T_1 spaces from T_0 spaces.

- Every T_2 (Hausdorff) space is also T_1. That is, if a space is Hausdorff, it automatically satisfies the T_1 condition.

- A T_1 space is not necessarily Hausdorff (i.e., T_2). For instance, the co-finite topology on an infinite set is T_1 but not Hausdorff.

- A T_1 space is closed under taking subspaces. That is, any subspace of a T_1 space is also T_1.

- A T_1 space is always a T_0 space. This follows because in a T_1 space, for every distinct pair of points, open sets exist that distinguish between them, which also satisfies the T_0 condition.

- If X is a finite space with the discrete topology, then X is a T_1 space. In fact, any finite set with any topology where every subset is open is a T_1 space.

Relation to Other Separation Axioms

1. T_1 spaces are stronger than T_0 spaces but weaker than T_2 (Hausdorff) spaces. While in a T_1 space, singletons are closed, it does not guarantee that any two distinct points have disjoint open neighborhoods, which is required for Hausdorff spaces.

2. The next level of separation is T_2, where distinct points must have disjoint open neighborhoods, a stronger condition than the existence of separating open sets as in T_1 spaces.

Applications of T_1 Spaces

1. The property that singletons are closed in a T_1 space is useful in various areas of topology, particularly in the study of compactness and convergence.

2. Many commonly studied spaces, including metric spaces (which are always T_1), and the real line $\mathbb{R}$ with the standard topology, are examples of T_1 spaces. This property is important for understanding continuous functions, compact sets, and separation of points in a space.

Conclusion

In conclusion, a T_1 space is a topological space in which singletons are closed sets, and distinct points can be separated by open sets. This property makes T_1 spaces more structured than T_0 spaces but less restrictive than T_2 (Hausdorff) spaces. Understanding the T_1 property is important for the study of separation axioms in topology, and it has numerous applications in various areas of mathematics.

3.3 T_2 Spaces (Hausdorff Spaces)

A topological space X is called a T_2 space or **Hausdorff space** if for every pair of distinct points $x, y \in X$, there exist disjoint neighborhoods of x and y. Hausdorff spaces are important because limits of sequences are unique.

Definition

A topological space X is said to be Hausdorff if for every pair of distinct points $x \neq y$, there exist open sets U and V such that:

$$x \in U, \quad y \in V, \quad \text{and} \quad U \cap V = \emptyset.$$

This means that any two distinct points in a Hausdorff space can be separated by disjoint open neighborhoods. The Hausdorff property ensures that limits of sequences (or nets) in these spaces are unique, which is a crucial property in many areas of topology and analysis.

Examples of Hausdorff Spaces

1. **Euclidean Space $\mathbb{R}^n$**: The Euclidean space with the standard topology is a Hausdorff space. For any two distinct points, say x and y, one can choose disjoint open balls centered at these points, thereby separating them. This follows from the fact that the Euclidean metric is continuous and allows the construction of such disjoint open sets.

2. **Metric Spaces**: Every metric space is Hausdorff. This is a direct consequence of the fact that in a metric space, for any two distinct points $x, y \in X$, there exists a positive distance between them. Using this distance, one can construct disjoint open balls around x and y. Specifically, for any two distinct points x and y, we can find radii r_x and r_y such that the open balls $B(x, r_x)$ and $B(y, r_y)$ are disjoint.

3. **Finite Discrete Space**: Any discrete topological space is Hausdorff. In the discrete topology, every singleton set $\{x\}$ is open, so for any two distinct points x and y, the sets $\{x\}$ and $\{y\}$ are disjoint open neighborhoods of x and y, respectively.

4. **The Real Line $\mathbb{R}$**: The real line $\mathbb{R}$ with the standard topology is Hausdorff. For any two distinct points $x, y \in \mathbb{R}$, we can always find disjoint open intervals containing them. This follows from the fact that open intervals in the real line are a basis for the topology, and they can be used to separate points.

5. **The Unit Circle $\mathbb{S}^1$**: The unit circle $\mathbb{S}^1 \subset \mathbb{R}^2$ with the subspace topology is Hausdorff. Since $\mathbb{R}^2$ is Hausdorff, the subspace topology on $\mathbb{S}^1$ inherits the Hausdorff property.

6. **Any Compact Hausdorff Space**: Any compact space that is Hausdorff retains the Hausdorff property. In fact, compact Hausdorff spaces are particularly important in topology and analysis due to their nice properties, such as the fact that every net or sequence has a convergent subsequence (Bolzano-Weierstrass property).

Properties of Hausdorff Spaces

1. **Uniqueness of Limits**: In a Hausdorff space, limits of sequences (or nets) are unique. This is a crucial property in analysis, as it ensures that a convergent sequence (or net) has exactly one limit point.

2. **Compactness and Hausdorff**: If a space is compact and Hausdorff, then it is also closed and bounded in Euclidean space. Furthermore, compact subsets of Hausdorff spaces are closed, and compact sets in Hausdorff spaces are also sequentially compact (every sequence has a convergent subsequence).

3. **Closed Diagonal**: In a Hausdorff space X, the diagonal $\Delta = \{(x, x) \mid x \in X\}$ is a closed set in the product space $X \times X$. This is important in the study of product topologies and in the theory of continuous functions.

4. **Separation of Points**: The Hausdorff condition guarantees that distinct points can be separated by disjoint open sets. This provides a strong form of separation between points and is crucial in many areas of topology and geometry.

5. **Subspaces of Hausdorff Spaces**: Any subspace of a Hausdorff space is also Hausdorff. This is an important property when working with subspaces, as it ensures that subspaces of Hausdorff spaces retain the Hausdorff property.

6. **Product of Hausdorff Spaces**: The product of two Hausdorff spaces is Hausdorff. This follows from the fact that in the product topology, we can separate points in the product space by taking disjoint open sets around the individual components.

Applications of Hausdorff Spaces

1. Hausdorff spaces are fundamental in the study of limits, continuity, and convergence. The uniqueness of limits in Hausdorff spaces makes

them a key concept in analysis, particularly in the study of topological spaces where limits play a central role.

2. In algebraic topology, Hausdorff spaces are often used to study continuous functions, compactness, and convergence. Many classical results in topology, such as the Tychonoff theorem (the product of compact Hausdorff spaces is compact), rely on the Hausdorff property.

3. In the context of metric spaces, the Hausdorff property ensures the uniqueness of the limit of a Cauchy sequence, which is essential for the completeness of the space.

Conclusion

A Hausdorff space is a topological space in which distinct points can be separated by disjoint open neighborhoods. This separation property leads to the uniqueness of limits, making Hausdorff spaces a cornerstone of analysis and topology. Many important spaces in mathematics, such as Euclidean spaces, metric spaces, and compact Hausdorff spaces, possess the Hausdorff property, which plays a crucial role in their study and applications.

3.4 Other Separation Properties

In addition to the basic T_0, T_1, and T_2 separation axioms, there are other important separation properties that arise in more advanced topology. These include T_3, T_4, completely regular spaces, and Tychonoff spaces, which strengthen the separation properties of the space.

3.4.1 T_3 Spaces (Regular Spaces)

A topological space is T_3 (regular) if it is T_1 and, for any point $x \in X$ and a closed set $C \subseteq X$ with $x \notin C$, there exist disjoint open sets U and V such that $x \in U$ and $C \subseteq V$. This means that not only can distinct points be separated by open sets, but any point can be separated from a closed set that does not contain it.

3.4.2 T_4 Spaces (Normal Spaces)

A topological space is T_4 (normal) if it is T_1 and, for any two disjoint closed sets A and B, there exist disjoint open sets U and V such that $A \subseteq U$ and $B \subseteq V$. The T_4 property is stronger than the T_3 property because it requires

Chapter 3: Separation Axioms

the ability to separate two disjoint closed sets, not just a point and a closed set.

Examples of Regular and Normal Spaces

1. **Euclidean Space** $\mathbb{R}^n$: The Euclidean space with the standard topology is both regular and normal. For any point and closed set, we can find disjoint open neighborhoods separating them, and similarly, for any two disjoint closed sets, disjoint open sets can be found.

2. **Metric Spaces**: Every metric space is both regular and normal. Since metric spaces are Hausdorff (thus T_2), they automatically satisfy the conditions for being regular and normal.

3. **Discrete Topology**: Any space with the discrete topology is trivially regular and normal. In the discrete topology, every subset is open, so disjoint open sets can always be found to separate points or disjoint closed sets.

3.4.3 Completely Regular and Tychonoff Spaces

A space is **completely regular** if it is T_1 and, for any closed set C and point $x \notin C$, there is a continuous function $f : X \to [0, 1]$ such that $f(x) = 0$ and $f(C) = 1$. This property ensures that a continuous function can be used to separate points and closed sets, which is a stronger form of regularity.

A **Tychonoff space** is a completely regular T_0 space. Tychonoff spaces are important in the study of spaces where the separation of points by continuous functions is a key aspect. These spaces are fundamental in the study of function spaces, compactness, and the theory of continuous mappings.

Examples of Completely Regular and Tychonoff Spaces

1. **Euclidean Space** $\mathbb{R}^n$: Euclidean spaces are completely regular because they are T_2 and satisfy the separation condition for continuous functions.

2. **Metric Spaces**: Since every metric space is T_2, it is automatically completely regular.

3. **The Sorgenfrey Line**: The Sorgenfrey line is a completely regular T_0 space, which is an example of a Tychonoff space that is not T_2. It can be separated using continuous functions, though it is not Hausdorff.

4. **Discrete Spaces**: Any discrete space is completely regular because every singleton set is closed, and continuous functions can easily be constructed for such spaces.

Properties of Regular, Normal, and Completely Regular Spaces

- **Regularity and Normality**: A space that is T_3 (regular) and T_4 (normal) has strong separation properties. Regular spaces allow separation between points and closed sets, while normal spaces allow separation of two disjoint closed sets.

- **Tychonoff Property**: A Tychonoff space, which is a completely regular T_0 space, has the property that points can be separated from closed sets by continuous functions, which is a crucial property in functional analysis and topology.

- **Continuous Functions in Regular Spaces**: In regular spaces, continuous functions can be used to separate points from closed sets. This is important in many areas of topology, particularly when dealing with the separation of points and closed sets in the context of continuous maps.

- **The Urysohn Lemma**: In normal spaces, the Urysohn Lemma provides a powerful tool for constructing continuous functions that separate disjoint closed sets, which is essential for understanding the structure of normal spaces.

Applications of Regular, Normal, and Completely Regular Spaces

1. Function Spaces: The concept of completely regular spaces is important in the study of function spaces and continuous functions. Completely regular spaces allow for the separation of points from closed sets by continuous functions, which is crucial in functional analysis.

2. Compactness: In normal spaces, compact subsets have important properties, such as being closed. This is useful when working with compactness in topological spaces.

3. General Topology: Regular and normal spaces are frequently encountered in general topology, where the ability to separate points or closed sets is a key property in many proofs and constructions.

4. Urysohn's Lemma: This lemma is a central result in normal spaces, allowing for the construction of continuous functions that separate disjoint closed sets, and is used in various fields, including algebraic topology and geometric topology.

Conclusion

The separation properties T_3 (regular), T_4 (normal), and completely regular (Tychonoff) spaces are important in topology because they allow for greater flexibility in separating points and closed sets by open sets or continuous functions. These properties provide the foundation for many key theorems in topology, including Urysohn's Lemma and results in functional analysis. Regular and normal spaces are crucial for understanding compactness and continuous functions, while Tychonoff spaces provide a more general setting for continuous separation by functions.

3.5 Relations Between Separation Axioms

The separation axioms form a hierarchy of increasingly stronger conditions:

$$T_4 \implies T_3 \implies T_2 \implies T_1 \implies T_0$$

Each implication is strict, meaning that there exist examples of spaces that satisfy one axiom but not the stronger ones. This hierarchy helps in understanding the relationships between different classes of topological spaces and how they can be classified based on their separation properties. For example, every T_4 space is T_3, T_2, T_1, and T_0, but not every T_2 space is T_3, and not every T_3 space is T_4.

The Hierarchy Explained

The hierarchy of separation axioms represents a range of topological spaces that vary in their ability to separate points, closed sets, and neighborhoods. Let us further examine the implications:

1. T_0 Spaces: The weakest separation axiom in the hierarchy, guaranteeing that for any pair of distinct points x and y, there is an open

set containing one but not the other. These spaces are useful in cases where the topological structure does not require a strong ability to separate points.

2. T_1 Spaces: This axiom introduces the condition that for every pair of distinct points, each has an open neighborhood not containing the other, making singletons closed. This axiom is often seen in spaces where the idea of separation is more intuitive, such as in spaces that describe certain kinds of discrete or algebraic structures.

3. T_2 (Hausdorff) Spaces: This is one of the most commonly encountered separation axioms, which requires that any two distinct points in the space can be separated by disjoint open sets. The Hausdorff property plays a significant role in ensuring the uniqueness of limits of sequences, which is a cornerstone in analysis and the study of convergence in topological spaces.

4. T_3 (Regular) Spaces: A T_3 space is not only T_1 but also ensures that any point and any closed set can be separated by disjoint open sets. This additional regularity condition is particularly useful in the study of the behavior of continuous functions and their interactions with closed sets.

5. T_4 (Normal) Spaces: The strongest of the separation axioms, where two disjoint closed sets can always be separated by disjoint open sets. This property is essential in classical results in topology, such as the Tietze Extension Theorem and Urysohn's Lemma, and it plays a significant role in the study of compactness and function spaces.

Strictness of the Implications

While the hierarchy suggests that stronger separation axioms lead to greater control over the topological structure of a space, the strictness of the implications is important to note. For example:

1. There exist T_2 spaces that are not T_3. The standard topology on the real line $\mathbb{R}$ is Hausdorff, but it is not regular in some pathological cases, like in spaces that are not first-countable.

2. Similarly, there are T_1 spaces that are not T_2, which can occur in cases where there is no guarantee of disjointness between neighborhoods of distinct points.

This strictness illustrates that the separation properties are not interchangeable but provide increasingly sophisticated structures that build on the simpler conditions.

3.6 Additional Separation Properties and Applications

In addition to the basic T_0, T_1, T_2, T_3, and T_4 separation axioms, there are other important separation properties that come into play in more specialized areas of topology. These further separate spaces based on the existence and structure of neighborhoods and their relationships to points and sets.

3.6.1 Completely Hausdorff Spaces

A space is called **completely Hausdorff** (also known as T_5) if for every pair of distinct points $x \neq y$, there exist disjoint open sets U and V such that $x \in U$ and $y \in V$, and the closure of U is disjoint from the closure of V. This axiom provides a stronger form of separation than the Hausdorff condition and guarantees a more refined distinction between points and their closed neighborhoods.

3.6.2 Completely Regular Spaces (Tychonoff Spaces)

A space is called **completely regular** if it is T_1 and for any closed set $C \subseteq X$ and point $x \notin C$, there exists a continuous function $f : X \to [0, 1]$ such that $f(x) = 0$ and $f(C) = 1$. A completely regular space can be thought of as one where points and closed sets are sufficiently "separated" by continuous functions, making them useful in the study of function spaces and the general theory of continuous mappings.

3.6.3 Applications of Separation Axioms

The separation properties in topology are not just abstract definitions; they have far-reaching implications in a variety of mathematical fields. Some notable applications include:

Analysis

In functional analysis, the Hausdorff property (T_2) is particularly important because it guarantees the uniqueness of limits of sequences. This property is foundational in defining convergent sequences and series, which are central to calculus and differential equations. Moreover, the compactness of subsets in Hausdorff spaces has significant implications in optimization and variational analysis.

Algebraic Topology

Separation axioms play a critical role in algebraic topology, where spaces are classified based on their homotopy properties. For example, the T_2 (Hausdorff) condition is crucial when working with the concept of separation of points in spaces, allowing us to define continuous functions and study their properties in terms of the connectedness and compactness of spaces.

Geometry

In differential geometry, the T_2 and T_3 separation axioms allow for the separation of geometric objects such as curves, surfaces, and manifolds, facilitating the study of their topology. The T_4 (normal) axiom is significant when discussing the extension of continuous functions over closed sets, which is essential in differential geometry.

Topology of Manifolds

The higher separation properties such as T_3 and T_4 are important when dealing with manifolds and the various topological properties of these spaces. For example, normality (T_4) plays a crucial role in the study of the partition of unity, a concept used in manifold theory to construct continuous functions with prescribed properties.

Model Theory and Logic

In model theory, separation properties help in understanding the structure of topological spaces as models for various logical systems. The distinction between different types of spaces, particularly the difference between T_0 and T_2, can be crucial when analyzing the behavior of logical formulas in topological settings.

Set Theory and Foundations of Mathematics

Set-theoretic topology, which studies the properties of spaces from the perspective of sets, uses separation axioms to classify spaces based on the distinguishability of points and sets. In this context, the study of T_0, T_1, and higher separation properties plays an important role in understanding the foundational aspects of topology, especially when discussing the notions of open sets, closure, and limit points.

3.6.4 Connections Between Separation Axioms and Other Topological Properties

In addition to their practical applications, separation axioms are closely related to other topological properties such as:

Compactness

In a Hausdorff space (T_2), compact sets are closed. This is a fundamental property in both functional analysis and general topology, ensuring that compact sets exhibit desirable topological characteristics, such as being bounded and having limit points that are contained within the set.

Connectedness

In T_2 spaces, connected sets have the property that they cannot be partitioned into disjoint open sets. This property is useful in determining the structure of spaces, especially when studying phenomena like path-connectedness and the classification of spaces based on their topological properties.

Lindelöf Property

A topological space is Lindelöf if every open cover has a countable subcover. The Lindelöf property is closely related to the separation axioms, especially T_1 and T_2, as the behavior of open covers in these spaces can help determine the space's compactness or metrizability.

Metrizability

A space is metrizable if it has a metric that induces its topology. Metrizable spaces are typically T_2 (Hausdorff), and they often satisfy additional separation axioms, such as regularity (T_3) or normality (T_4). The metrizability criterion is an important tool for analyzing the structure of topological spaces and understanding their geometric properties.

MCQs: Separation Axioms

1. A topological space (X, τ) is said to be T_0 if:

    (a) For any two distinct points in X, there exists an open set containing one of the points but not the other.

    (b) For any two distinct points in X, there exists an open set containing both points.

    (c) For any two distinct points in X, there exists an open set containing both points but not their union.

(d) For any two distinct points in X, there exists an open set containing one of the points and another open set containing the other point.

Answer: (a)
Explanation: A space is T_0 if for any two distinct points, there is an open set that contains one of the points and not the other.

2. In a T_1 space, also known as a Frechet space, which of the following statements is true?

   (a) Every pair of distinct points can be separated by disjoint open sets.

   (b) Every pair of distinct points can be separated by open sets such that each point is contained in one of these open sets.

   (c) Every point in the space is an isolated point.

   (d) Every point can be separated from every other point by open sets.

Answer: (b)
Explanation: In a T_1 space, each point is a closed set, meaning every pair of distinct points can be separated by open sets, but not necessarily disjoint open sets.

3. A topological space (X, τ) is T_2 (Hausdorff) if:

   (a) Any two distinct points can be separated by disjoint open sets.

   (b) Any two distinct points can be separated by open sets where the intersection of these sets is empty.

   (c) Any two distinct points have neighborhoods that do not intersect.

   (d) Any two distinct points have neighborhoods that are not disjoint.

Answer: (a)
Explanation: A space is T_2 (Hausdorff) if for any two distinct points, there exist disjoint open sets containing each point.

4. Which of the following separation axioms implies the others?

   (a) T_0

   (b) T_1

(c) T_2

(d) T_1 and T_2

Answer: (c)
Explanation: A T_2 (Hausdorff) space implies T_1 and T_0 since the Hausdorff condition is stronger and encompasses the conditions required for T_0 and T_1.

5. In a T_0 space, the closure of a point x is:

    (a) The union of all open sets containing x.

    (b) The set of all points that cannot be separated from x by open sets.

    (c) The set of all points that are not contained in any open set containing x.

    (d) The intersection of all open sets containing x.

Answer: (b)
Explanation: In a T_0 space, the closure of a point x is the set of all points that cannot be separated from x by open sets.

6. Which separation axiom ensures that every singleton set is closed?

    (a) T_0

    (b) T_1

    (c) T_2

    (d) T_3

Answer: (b)
Explanation: A T_1 space, or Frechet space, ensures that every singleton set is closed.

7. In a T_2 space, what is true about compact subsets?

    (a) Every compact subset is closed.

    (b) Every compact subset is open.

    (c) Every compact subset is both open and closed.

    (d) Compact subsets are neither open nor closed.

Answer: (a)
Explanation: In a T_2 (Hausdorff) space, compact subsets are closed.

8. A space X is T_3 (regular) if:

    (a) It is T_1 and for every closed set C and a point x not in C, there are disjoint open sets containing x and C.

    (b) It is T_2 and for every pair of disjoint open sets, there is a point in each set.

    (c) It is T_1 and every point can be separated from a closed set by open sets.

    (d) It is T_2 and every point can be separated from a closed set by open sets.

Answer: (a)
Explanation: A space is T_3 if it is T_1 and regular, meaning that points and closed sets can be separated by disjoint open sets.

9. In a T_4 (normal) space:

    (a) Every pair of disjoint closed sets can be separated by disjoint open neighborhoods.

    (b) Every pair of disjoint open sets can be separated by disjoint closed neighborhoods.

    (c) Every pair of disjoint closed sets can be separated by open sets.

    (d) Every pair of disjoint open sets can be separated by closed sets.

Answer: (a)
Explanation: In a T_4 (normal) space, any two disjoint closed sets can be separated by disjoint open neighborhoods.

10. If a space is T_4 (normal) and compact, what other property does it have?

    (a) It is T_5 (completely normal).

    (b) It is T_3 (regular).

    (c) It is T_2 (Hausdorff).

    (d) It is T_1.

Chapter 3: Separation Axioms 123

Answer: (a)
Explanation: In a compact T_4 (normal) space, it is also T_5 (completely normal), meaning it satisfies all properties of T_4 and more.

11. In a T_2 space, which of the following is true about sequences?

    (a) Every convergent sequence converges to a unique limit.

    (b) Every convergent sequence may converge to more than one limit.

    (c) There are sequences that do not converge.

    (d) All sequences converge.

    Answer: (a)
    Explanation: In a T_2 (Hausdorff) space, every convergent sequence converges to a unique limit due to the separation property.

12. Which separation axiom is also known as the Hausdorff property?

    (a) T_0

    (b) T_1

    (c) T_2

    (d) T_3

    Answer: (c)
    Explanation: The Hausdorff property is synonymous with the T_2 separation axiom.

13. In a T_1 space, the intersection of two closed sets is:

    (a) Always closed.

    (b) Always open.

    (c) Not necessarily closed.

    (d) Not necessarily open.

    Answer: (a)
    Explanation: In a T_1 space, the intersection of closed sets is always closed.

14. A T_2 space is also known as:

(a) A regular space.

(b) A normal space.

(c) A Hausdorff space.

(d) A completely normal space.

Answer: (c)
Explanation: A T_2 space is also known as a Hausdorff space, where every pair of distinct points can be separated by disjoint open sets.

15. Which of the following properties is stronger than T_1 but weaker than T_2?

    (a) T_0

    (b) T_3

    (c) T_4

    (d) T_5

Answer: (b)
Explanation: The T_3 property is stronger than T_1 but weaker than T_2.

16. A space that is T_4 is also T_3. Which of the following is also true for a T_4 space?

    (a) It is normal.

    (b) It is regular.

    (c) It is compact.

    (d) It is connected.

Answer: (a)
Explanation: A T_4 (normal) space is also T_3 (regular) and has the property of being normal.

17. In a T_2 space, what happens to compact sets?

    (a) They are always closed.

    (b) They are always open.

    (c) They are sometimes open.

Chapter 3: Separation Axioms

(d) They are always dense.

Answer: (a)
Explanation: In a T_2 (Hausdorff) space, compact sets are always closed.

18. In a T_0 space, which of the following statements is false?

    (a) Not all points can be separated by open sets.
    (b) Any two distinct points can be separated by disjoint open sets.
    (c) Every point can be isolated by open sets.
    (d) There may exist points that cannot be separated by open sets.

    Answer: (b)
    Explanation: In a T_0 space, not all points can be separated by disjoint open sets; rather, there must be some open set that contains one point and not the other.

19. Which property is a stronger version of T_2?

    (a) T_3
    (b) T_4
    (c) T_5
    (d) T_1

    Answer: (b)
    Explanation: T_4 (normal) is a stronger version of T_2 (Hausdorff), as it also ensures the separation of disjoint closed sets.

20. In a T_2 space, which of the following is true?

    (a) Every sequence converges to a unique limit.
    (b) Every open set is closed.
    (c) Every compact set is open.
    (d) Every compact set is closed.

    Answer: (d)
    Explanation: In a T_2 (Hausdorff) space, every compact set is closed.

21. If a space is T_3, which property does it necessarily satisfy?

    (a) T_2

    (b) T_4

    (c) T_0

    (d) T_1

    Answer: (a)
    Explanation: A T_3 space (regular) necessarily satisfies T_2 (Hausdorff) since regularity implies the separation of points and closed sets, which is stronger than T_2.

22. Which separation axiom implies that every compact subset is closed?

    (a) T_0

    (b) T_1

    (c) T_2

    (d) T_4

    Answer: (c)
    Explanation: A T_2, when compact, ensures that compact subsets are closed.

23. In a T_0 space, if x and y are distinct points, which of the following must be true?

    (a) There is an open set containing x but not y, or an open set containing y but not x.

    (b) There is an open set containing both x and y.

    (c) Every open set contains x and y.

    (d) There are open sets containing x and y such that their intersection is empty.

    Answer: (a)
    Explanation: In a T_0 space, for distinct points x and y, there exists an open set containing one but not the other.

24. If a space is T_1, which property does it automatically have?

(a) T_0

(b) T_2

(c) T_3

(d) T_4

Answer: (a)
Explanation: A T_1 space (Frechet) automatically satisfies T_0 since T_1 is a stronger condition.

25. In a T_2 space, what can be said about the closure of a set?

    (a) It is always open.

    (b) It is always closed.

    (c) It is sometimes open and sometimes closed.

    (d) It is always dense.

    Answer: (b)
    Explanation: In a T_2 (Hausdorff) space, the closure of a set is always closed.

26. In a T_2 space, what property do compact subsets have?

    (a) They are always open.

    (b) They are always closed.

    (c) They are sometimes open.

    (d) They are sometimes closed.

    Answer: (b)
    Explanation: In a T_2 (Hausdorff) space, compact subsets are always closed.

27. Which separation axiom is equivalent to saying that every compact subset is closed in a Hausdorff space?

    (a) T_1

    (b) T_2

    (c) T_3

(d) T_4

Answer: (b)
Explanation: In a T_2 (Hausdorff) space, compact subsets are closed, so this property is equivalent to T_2.

28. Which axiom guarantees the separation of any two disjoint closed sets by open neighborhoods?

    (a) T_1

    (b) T_2

    (c) T_3

    (d) T_4

Answer: (d)
Explanation: The T_4 (normal) axiom guarantees that any two disjoint closed sets can be separated by disjoint open neighborhoods.

29. Which separation property is known as the "Frechet" condition?

    (a) T_0

    (b) T_1

    (c) T_2

    (d) T_3

Answer: (b)
Explanation: The T_1 property is also known as the Frechet condition.

30. A space that satisfies T_4 (normal) and T_2 (Hausdorff) is known as:

    (a) Completely normal.

    (b) Regular.

    (c) Compact.

    (d) Connected.

Answer: (a)
Explanation: A space that is T_4 (normal) and T_2 (Hausdorff) is known as completely normal.

Chapter 3: Separation Axioms

31. In a T_2 space, any two distinct points can be separated by:

    (a) Disjoint open sets.
    (b) Disjoint closed sets.
    (c) Open sets with empty intersection.
    (d) Closed sets with empty intersection.

    Answer: (a)
    Explanation: In a T_2 (Hausdorff) space, any two distinct points can be separated by disjoint open sets.

32. In a T_1 space, any point can be separated from a closed set by:

    (a) Disjoint open sets.
    (b) Open sets that intersect.
    (c) Closed sets that intersect.
    (d) Closed sets with empty intersection.

    Answer: (a)
    Explanation: In a T_1 space, any point can be separated from a closed set by disjoint open sets.

33. A space where every pair of distinct points can be separated by disjoint open sets is:

    (a) T_0
    (b) T_1
    (c) T_2
    (d) T_3

    Answer: (c)
    Explanation: A space where every pair of distinct points can be separated by disjoint open sets is T_2 (Hausdorff).

34. If a space is T_3 and T_2, what additional property does it have?

    (a) It is T_4 (normal).
    (b) It is T_0.

(c) It is T_1.

(d) It is T_5 (completely normal).

Answer: (c)
Explanation: A space that is both T_3 (regular) and T_2 (Hausdorff) is T_1.

35. Which axiom is known for ensuring that any two disjoint closed sets can be separated by open neighborhoods?

    (a) T_1

    (b) T_2

    (c) T_3

    (d) T_4

    Answer: (d)
    Explanation: The T_4 (normal) axiom ensures that any two disjoint closed sets can be separated by disjoint open neighborhoods.

36. In a T_1 space, any finite set is:

    (a) Closed.

    (b) Open.

    (c) Compact.

    (d) Dense.

    Answer: (a)
    Explanation: In a T_1 space, any finite set is closed.

37. If a space is T_4 (normal) and compact, what is it also?

    (a) T_5 (completely normal).

    (b) T_2 (Hausdorff).

    (c) T_3 (regular).

    (d) T_1.

    Answer: (a)
    Explanation: A compact T_4 (normal) space is also T_5 (completely normal).

Chapter 3: Separation Axioms

38. A space is T_3 (regular) if:

    (a) It is T_2 and every point can be separated from a closed set by open sets.

    (b) It is T_1 and every point can be separated from a closed set by disjoint open sets.

    (c) It is T_2 and every pair of disjoint closed sets can be separated by disjoint open neighborhoods.

    (d) It is T_0 and every closed set can be separated from any point by open sets.

    Answer: (a)
    Explanation: A space is T_3 (regular) if it is T_2 (Hausdorff) and every point can be separated from a closed set by open sets.

39. Which property ensures that a space is both T_2 and compact?

    (a) T_1

    (b) T_2

    (c) T_4

    (d) T_5

    Answer: (d)
    Explanation: A space that is T_2 (Hausdorff) and compact is T_5 (completely normal).

40. In a T_0 space, the closure of a set A is:

    (a) The intersection of all open sets containing A.

    (b) The union of all open sets containing A.

    (c) The set of all points that are not in A.

    (d) The complement of A.

    Answer: (a)
    Explanation: In a T_0 space, the closure of a set A is the intersection of all open sets containing A.

41. In a T_1 space, which of the following is true about points and closed sets?

(a) Points can be separated from closed sets by open sets.

(b) Points and closed sets cannot be separated.

(c) Points and closed sets must intersect.

(d) Points and closed sets are always disjoint.

Answer: (a)
Explanation: In a T_1 space, points can be separated from closed sets by open sets.

42. If a space is T_4 (normal) and T_1, it is also:

    (a) T_2

    (b) T_3

    (c) T_5

    (d) T_0

    Answer: (a)
    Explanation: A T_4 (normal) and T_1 space is also T_2 (Hausdorff).

43. In a T_2 space, what can be said about the intersection of two compact sets?

    (a) It is always compact.

    (b) It is always open.

    (c) It is sometimes open.

    (d) It is always dense.

    Answer: (a)
    Explanation: In a T_2 (Hausdorff) space, the intersection of two compact sets is always compact.

44. A space where every pair of distinct points can be separated by disjoint open sets is known as:

    (a) T_0

    (b) T_1

    (c) T_2

Chapter 3: Separation Axioms

(d) T_3

Answer: (c)
Explanation: A space where every pair of distinct points can be separated by disjoint open sets is known as T_2 (Hausdorff).

45. In a T_4 space, the intersection of any two disjoint closed sets can be:

    (a) Separated by open sets.

    (b) Separated by closed sets.

    (c) Separated by dense sets.

    (d) Separated by compact sets.

Answer: (a)
Explanation: In a T_4 (normal) space, any two disjoint closed sets can be separated by disjoint open sets.

46. In a T_2 space, what happens to the union of two compact sets?

    (a) It is always compact.

    (b) It is always closed.

    (c) It is sometimes compact.

    (d) It is always dense.

Answer: (b)
Explanation: In a T_2 (Hausdorff) space, the union of two compact sets is not necessarily compact but is always closed.

47. In a T_1 space, which of the following statements is true?

    (a) All points can be separated by disjoint open sets.

    (b) Every compact subset is open.

    (c) Every finite subset is closed.

    (d) Every open set is closed.

Answer: (c)
Explanation: In a T_1 space, all finite subsets are closed.

48. Which of the following properties does not necessarily follow from a space being T_2?

    (a) Every compact subset is closed.
    (b) Every sequence has at most one limit point.
    (c) Every point is isolated.
    (d) Every compact subset is compact.

    Answer: (c)
    Explanation: In a T_2 (Hausdorff) space, not every point is isolated; only compact subsets are guaranteed to be closed.

49. Which separation axiom ensures that every compact set in a space is closed and every two disjoint closed sets can be separated by open neighborhoods?

    (a) T_1
    (b) T_2
    (c) T_3
    (d) T_4

    Answer: (d)
    Explanation: The T_4 (normal) axiom ensures that every compact set is closed and disjoint closed sets can be separated by open neighborhoods.

50. Which axiom ensures that every pair of distinct points can be separated by open neighborhoods?

    (a) T_0
    (b) T_1
    (c) T_2
    (d) T_3

    Answer: (c)
    Explanation: The T_2 (Hausdorff) axiom ensures that every pair of distinct points can be separated by disjoint open neighborhoods.

51. In a T_3 (regular) space, what happens to the intersection of two closed sets?

(a) It is always compact.

(b) It is always open.

(c) It is sometimes closed.

(d) It is always closed.

Answer: (d)
Explanation: In a T_3 (regular) space, the intersection of two closed sets is always closed.

52. Which of the following properties implies T_4 (normal)?

    (a) T_2

    (b) T_3

    (c) T_5

    (d) T_0

Answer: (b)
Explanation: T_3 (regular) spaces are T_4 (normal) if they are also T_2 (Hausdorff).

53. In a T_1 space, which of the following is necessarily true?

    (a) Every point is isolated.

    (b) Every closed set is compact.

    (c) Every singleton set is closed.

    (d) Every subset is closed.

Answer: (c)
Explanation: In a T_1 space, every singleton set is closed.

54. Which of the following spaces must be T_2 (Hausdorff) if it is T_5 (completely normal)?

    (a) T_0

    (b) T_1

    (c) T_2

    (d) T_3

Answer: (b)
Explanation: A T_5 (completely normal) space must be T_1

55. If a space is T_2 and compact, what additional property does it automatically have?

    (a) It is T_4 (normal).
    (b) It is T_3 (regular).
    (c) It is T_0.
    (d) It is T_1.

Answer: (a)
Explanation: A compact T_2 (Hausdorff) space is automatically T_4 (normal).

56. In a T_0 space, can you always find disjoint open sets for every pair of distinct points?

    (a) Yes, for every pair of distinct points.
    (b) No, not always.
    (c) Yes, but only for some pairs.
    (d) Only if the space is compact.

Answer: (b)
Explanation: In a T_0 space, it is not always possible to find disjoint open sets for every pair of distinct points.

57. Which of the following is not necessarily a property of a T_1 space?

    (a) Every finite set is closed.
    (b) Every singleton is closed.
    (c) Every compact set is closed.
    (d) Every open set is closed.

Answer: (d)
Explanation: In a T_1 space, not every open set is necessarily closed.

58. If a space is T_4 and T_2, it is also known as:

(a) Completely normal.

(b) Regular.

(c) Compact.

(d) Connected.

Answer: (a)
Explanation: A space that is T_4 (normal) and T_2 (Hausdorff) is known as completely normal.

59. In a T_2 space, the union of two disjoint compact sets is:

    (a) Always compact.

    (b) Always closed.

    (c) Sometimes compact.

    (d) Sometimes closed.

    Answer: (b)
    Explanation: In a T_2 (Hausdorff) space, the union of two disjoint compact sets is always closed but not necessarily compact.

60. If a space is T_3 (regular) and T_2, what is it also?

    (a) T_4 (normal).

    (b) T_5 (completely normal).

    (c) T_0.

    (d) T_1.

    Answer: (d)
    Explanation: A space that is both T_3 (regular) and T_2 (Hausdorff) is T_1.

61. In a T_1 space, which of the following statements is false?

    (a) Every singleton set is closed.

    (b) Every finite set is closed.

    (c) Every infinite set is closed.

    (d) Every compact set is closed.

Answer: (c)

Explanation: In a T_1 space, infinite sets are not necessarily closed; only finite sets and singletons are guaranteed to be closed.

62. In a T_0 space, which property is guaranteed?

    (a) Every point is isolated.

    (b) Every two distinct points can be separated by disjoint open sets.

    (c) Every point can be distinguished from every closed set.

    (d) Every point can be distinguished from every other point by open sets.

Answer: (d)

Explanation: In a T_0 space, every pair of distinct points can be separated by open sets, but not necessarily disjoint open sets.

63. If a space is T_2 (Hausdorff) and T_5 (completely normal), it is also:

    (a) Compact.

    (b) Regular.

    (c) Normal.

    (d) Connected.

Answer: (c)

Explanation: A T_2 (Hausdorff) and T_5 (completely normal) space is also Normal.

64. In a T_2 space, what can be said about the closure of an open set?

    (a) It is always closed.

    (b) It is always open.

    (c) It is sometimes closed.

    (d) It is sometimes open.

Answer: (a)

Explanation: In a T_2 (Hausdorff) space, the closure of an open set is always closed.

Chapter 3: Separation Axioms 139

65. A space where every point can be separated from a closed set by disjoint open sets is known as:

    (a) T_1
    (b) T_2
    (c) T_3
    (d) T_4

    Answer: (c)
    Explanation: A space where every point can be separated from a closed set by disjoint open sets is T_3 (regular).

66. Which property ensures that any two disjoint closed sets can be separated by closed neighborhoods?

    (a) T_0
    (b) T_1
    (c) T_2
    (d) T_4

    Answer: (d)
    Explanation: The T_4 (normal) axiom ensures that any two disjoint closed sets can be separated by disjoint open neighborhoods, and it implies separation by closed neighborhoods in T_4 (normal) spaces.

67. In a T_3 (regular) space, the union of two closed sets is:

    (a) Always closed.
    (b) Always open.
    (c) Sometimes open.
    (d) Sometimes closed.

    Answer: (a)
    Explanation: In a T_3 (regular) space, the union of two closed sets is always closed.

68. A space is T_5 (completely normal) if and only if it is:

    (a) T_2 and T_4.

(b) T_1 and T_3.

(c) T_2 and T_1.

(d) T_4 and T_3.

Answer: (a)
Explanation: A space is T_5 (completely normal) if and only if it is T_2 (Hausdorff) and T_4 (normal).

69. In a T_2 (Hausdorff) space, which of the following is always true?

    (a) The intersection of two compact sets is compact.

    (b) The union of two compact sets is compact.

    (c) Every closed set is compact.

    (d) Every open set is compact.

Answer: (a)
Explanation: In a T_2 (Hausdorff) space, the intersection of two compact sets is always compact.

70. Which property is not guaranteed by T_1 but is ensured by T_2?

    (a) Separation of points by disjoint open sets.

    (b) Separation of closed sets by disjoint open sets.

    (c) Separation of points from closed sets by open sets.

    (d) Compact sets being closed.

Answer: (b)
Explanation: T_2 (Hausdorff) spaces guarantee that disjoint closed sets can be separated by disjoint open sets, whereas T_1 spaces do not necessarily provide this property.

71. In a T_1 space, which of the following is always true about finite sets?

    (a) They are closed.

    (b) They are open.

    (c) They are compact.

    (d) They are dense.

Chapter 3: Separation Axioms

Answer: (a)
Explanation: In a T_1 space, every finite set is closed.

72. Which separation property is required to ensure that any two disjoint closed sets can be separated by disjoint open sets?

    (a) T_0
    (b) T_1
    (c) T_2
    (d) T_4

    Answer: (d)
    Explanation: The T_4 (normal) axiom ensures that any two disjoint closed sets can be separated by disjoint open sets.

73. In a T_2 (Hausdorff) space, the closure of an intersection of compact sets is:

    (a) Always compact.
    (b) Always open.
    (c) Sometimes compact.
    (d) Sometimes closed.

    Answer: (a)
    Explanation: In a T_2 (Hausdorff) space, the closure of the intersection of compact sets is always compact.

74. If a space is T_2 (Hausdorff) and T_4 (normal), it is also:

    (a) Compact.
    (b) Regular.
    (c) Completely normal.
    (d) Connected.

    Answer: (c)
    Explanation: A T_2 (Hausdorff) and T_4 (normal) space is also T_5 (completely normal).

75. In a T_0 space, the closure of a set is always:

(a) Open.

(b) Closed.

(c) Dense.

(d) Empty.

Answer: (b)
Explanation: In a T_0 space, the closure of a set is always closed.

76. A topological space is T_0 (Kolmogorov) if:

    (a) For any two distinct points, there exists an open set containing one but not the other

    (b) Any two distinct points can be separated by disjoint open sets

    (c) It is compact

    (d) It is connected

Answer: (a)
Explanation: The T_0 axiom requires that for any $x \neq y$, there is an open set containing exactly one of them.

77. A topological space is T_1 if:

    (a) For any two distinct points, each has an open set not containing the other

    (b) Any two distinct points have disjoint open neighborhoods

    (c) It is compact

    (d) It is normal

Answer: (a)
Explanation: A T_1 space ensures that for any $x \neq y$, there exist open sets $U \ni x$, $V \ni y$ such that $y \notin U$, $x \notin V$.

78. A topological space is T_2 (Hausdorff) if:

    (a) Any two distinct points have disjoint open neighborhoods

    (b) For any point and closed set not containing it, there is an open set separating them

Chapter 3: Separation Axioms

(c) It is compact

(d) It is connected

Answer: (a)

Explanation: A Hausdorff space requires that for any $x \neq y$, there exist disjoint open sets $U \ni x, V \ni y$.

79. Every T_2 (Hausdorff) space is:

    (a) T_1

    (b) Not necessarily T_1

    (c) Compact

    (d) Normal

    Answer: (a)

    Explanation: The Hausdorff property (T_2) implies T_1, as disjoint open sets provide the required separation for T_1.

80. A topological space is T_1 if and only if:

    (a) Every singleton set is closed

    (b) Every singleton set is open

    (c) The space is compact

    (d) The space is connected

    Answer: (a)

    Explanation: A space is T_1 if for any point x, the set $X \setminus \{x\}$ is open, meaning $\{x\}$ is closed.

81. Every metric space is:

    (a) Hausdorff

    (b) Not necessarily Hausdorff

    (c) Compact

    (d) Connected

    Answer: (a)

    Explanation: In a metric space, distinct points can be separated by disjoint open balls, making it Hausdorff.

82. The discrete topology on any set is:

    (a) Hausdorff
    (b) Not Hausdorff
    (c) Not T_1
    (d) Not T_0

    Answer: (a)
    Explanation: In the discrete topology, every subset is open, so any two points can be separated by disjoint open sets.

83. The trivial topology on a set with at least two points is:

    (a) Not T_0
    (b) T_0
    (c) Hausdorff
    (d) Normal

    Answer: (a)
    Explanation: The trivial topology has only $\emptyset$ and X as open sets, so distinct points cannot be separated.

84. A topological space is regular if:

    (a) It is T_1 and for any point and closed set not containing it, there exist disjoint open sets separating them
    (b) It is Hausdorff and compact
    (c) It is T_1 and connected
    (d) It is normal

    Answer: (a)
    Explanation: Regularity requires T_1 and separation of a point from a closed set by disjoint open sets.

85. Every regular space is:

    (a) Hausdorff
    (b) Not necessarily Hausdorff

Chapter 3: Separation Axioms

(c) Compact

(d) Connected

Answer: (a)
Explanation: Regularity includes T_1, and separating points from closed sets allows separation of points, implying Hausdorff.

86. A topological space is normal if:

    (a) It is T_1 and any two disjoint closed sets can be separated by disjoint open sets

    (b) It is Hausdorff and compact

    (c) It is regular and connected

    (d) It is T_1 and separable

Answer: (a)
Explanation: Normality requires T_1 and separation of disjoint closed sets by disjoint open sets.

87. Every normal space is:

    (a) Regular

    (b) Not necessarily regular

    (c) Compact

    (d) Connected

Answer: (a)
Explanation: Normality implies regularity, as separating closed sets includes separating a point from a closed set.

88. Every compact Hausdorff space is:

    (a) Normal

    (b) Not necessarily normal

    (c) Connected

    (d) Separable

Answer: (a)

Explanation: In a compact Hausdorff space, disjoint closed (hence compact) sets can be separated by open sets, implying normality.

89. The real line $\mathbb{R}$ with the standard topology is:

    (a) Normal

    (b) Not normal

    (c) Not Hausdorff

    (d) Not T_1

    Answer: (a)
    Explanation: $\mathbb{R}$ is a metric space, hence Hausdorff and normal.

90. The co-finite topology on an infinite set is:

    (a) T_1

    (b) Not T_1

    (c) Normal

    (d) Not T_0

    Answer: (a)
    Explanation: Singletons are closed (their complements are co-finite and open), so the space is T_1.

91. A space is completely regular if:

    (a) It is T_1 and for any point and closed set not containing it, there exists a continuous function separating them

    (b) It is normal

    (c) It is compact

    (d) It is connected

    Answer: (a)
    Explanation: Complete regularity requires T_1 and separation by a continuous function to $[0, 1]$.

92. Every completely regular space is:

Chapter 3: Separation Axioms

(a) Regular

(b) Not necessarily regular

(c) Normal

(d) Compact

Answer: (a)
Explanation: Complete regularity implies regularity, as continuous functions can define open sets for separation.

93. A space is $T_{3\frac{1}{2}}$ (Tychonoff) if:

    (a) It is completely regular

    (b) It is normal

    (c) It is compact

    (d) It is Hausdorff

    Answer: (a)
    Explanation: The $T_{3\frac{1}{2}}$ axiom is equivalent to complete regularity.

94. Every metric space is:

    (a) Completely regular

    (b) Not completely regular

    (c) Compact

    (d) Connected

    Answer: (a)
    Explanation: Metric spaces allow continuous functions (e.g., distance functions) to separate points and closed sets.

95. Urysohn's lemma states that in a normal space, for any two disjoint closed sets A and B:

    (a) There exists a continuous function $f : X \to [0, 1]$ with $f(A) = 0$, $f(B) = 1$

    (b) They can be separated by open sets

    (c) They are compact

(d) They are connected

Answer: (a)
Explanation: Urysohn's lemma guarantees a continuous function separating disjoint closed sets in a normal space.

96. Tietze's extension theorem states that in a normal space, a continuous function $f : A \to \mathbb{R}$ on a closed subset A:

    (a) Can be extended to a continuous function on X

    (b) Is constant

    (c) Is bounded

    (d) Is open

Answer: (a)
Explanation: Tietze's theorem ensures that continuous functions on closed subsets extend to the entire space.

97. The co-countable topology on an uncountable set is:

    (a) T_1

    (b) Not T_1

    (c) Normal

    (d) Not T_0

Answer: (a)
Explanation: Singletons are closed (complements are countable), so the space is T_1.

98. A space with the co-finite topology on an infinite set is:

    (a) Not normal

    (b) Normal

    (c) Not Hausdorff

    (d) Not regular

Answer: (a)
Explanation: The co-finite topology is T_1 but not Hausdorff, hence not normal, as disjoint closed sets cannot be separated effectively.

Chapter 3: Separation Axioms

99. A space with the co-finite topology is:

    (a) Hausdorff

    (b) Not Hausdorff

    (c) Normal

    (d) Completely regular

    Answer: (b)
    Explanation: The co-finite topology is T_1, but open sets intersect non-trivially, so it is not Hausdorff.

100. The product of two Hausdorff spaces is:

    (a) Hausdorff

    (b) Not necessarily Hausdorff

    (c) Compact

    (d) Normal

    Answer: (a)
    Explanation: The product topology allows separation of points by products of open sets, preserving the Hausdorff property.

101. The product of two normal spaces is:

    (a) Not necessarily normal

    (b) Normal

    (c) Compact

    (d) Connected

    Answer: (a)
    Explanation: Normality is not preserved in products; e.g., $\mathbb{R} \times \mathbb{R}$ may fail normality in certain cases.

102. Every subspace of a Hausdorff space is:

    (a) Hausdorff

    (b) Not necessarily Hausdorff

    (c) Normal

(d) Compact

Answer: (a)
Explanation: The subspace topology inherits the Hausdorff property via intersections of open sets.

103. Every closed subspace of a normal space is:

    (a) Normal

    (b) Not necessarily normal

    (c) Hausdorff

    (d) Compact

    Answer: (a)
    Explanation: A closed subset of a normal space inherits normality, as closed sets can be separated.

104. A space is T_3 if:

    (a) It is regular

    (b) It is normal

    (c) It is Hausdorff

    (d) It is compact

    Answer: (a)
    Explanation: The T_3 axiom is equivalent to regularity (T_1 and point-closed set separation).

105. Every T_4 space is:

    (a) T_3

    (b) Not necessarily T_3

    (c) Compact

    (d) Connected

    Answer: (a)
    Explanation: A T_4 space (normal) is regular, hence T_3.

106. The space $\mathbb{Q} \subset \mathbb{R}$ with the subspace topology is:

Chapter 3: Separation Axioms 151

(a) Hausdorff

(b) Not Hausdorff

(c) Not T_1

(d) Not T_0

Answer: (a)
Explanation: $\mathbb{R}$ is Hausdorff, so its subspace $\mathbb{Q}$ is Hausdorff.

107. The Sorgenfrey line (lower limit topology on $\mathbb{R}$) is:

    (a) Hausdorff

    (b) Not Hausdorff

    (c) Not T_1

    (d) Not T_0

    Answer: (a)
    Explanation: The Sorgenfrey line, with basis $[a, b)$, is Hausdorff, as points can be separated by half-open intervals.

108. The Sorgenfrey line is:

    (a) Normal

    (b) Not normal

    (c) Compact

    (d) Connected

    Answer: (a)
    Explanation: The Sorgenfrey line is normal, as it is regular and satisfies separation properties for closed sets.

109. A space with the finite complement topology is:

    (a) T_1

    (b) Not T_1

    (c) Normal

    (d) Completely regular

Answer: (a)

Explanation: The finite complement topology is T_1, as singletons are closed.

110. The order topology on $\mathbb{Z}$ is:

    (a) Discrete

    (b) Not discrete

    (c) Not T_1

    (d) Not Hausdorff

    Answer: (a)

    Explanation: The order topology on $\mathbb{Z}$ has singletons as open sets, making it discrete and thus Hausdorff.

111. Every compact space is:

    (a) Normal

    (b) Not necessarily normal

    (c) Hausdorff

    (d) Completely regular

    Answer: (b)

    Explanation: Compactness does not imply normality without additional properties like Hausdorff.

112. A space is T_4 if:

    (a) It is normal

    (b) It is regular

    (c) It is Hausdorff

    (d) It is compact

    Answer: (a)

    Explanation: The T_4 axiom is equivalent to normality (T_1 and closed set separation).

113. The Niemytzki plane is:

Chapter 3: Separation Axioms 153

(a) Not normal

(b) Normal

(c) Compact

(d) Connected

Answer: (a)
Explanation: The Niemytzki (Moore) plane is Hausdorff but not normal, as certain closed sets cannot be separated.

114. Every second-countable Hausdorff space is:

    (a) Normal

    (b) Not necessarily normal

    (c) Compact

    (d) Connected

    Answer: (a)
    Explanation: Second-countable Hausdorff spaces are normal, as they satisfy Urysohn's lemma conditions.

115. The product $\mathbb{R} \times \mathbb{R}$ with the standard topology is:

    (a) Normal

    (b) Not normal

    (c) Not Hausdorff

    (d) Not T_1

    Answer: (a)
    Explanation: As a metric space, $\mathbb{R} \times \mathbb{R}$ is normal.

116. A space is regular if and only if:

    (a) For every point and open set containing it, there is an open set whose closure is contained in the original open set

    (b) It is normal

    (c) It is compact

    (d) It is connected

Answer: (a)

Explanation: This is an equivalent characterization of regularity, ensuring separation via closures.

117. The co-countable topology on an uncountable set is:

    (a) Not Hausdorff
    (b) Hausdorff
    (c) Normal
    (d) Completely regular

    Answer: (a)

    Explanation: The co-countable topology is T_1, but open sets intersect non-trivially, so it is not Hausdorff.

118. Every Lindelöf Hausdorff space is:

    (a) Normal
    (b) Not necessarily normal
    (c) Compact
    (d) Connected

    Answer: (b)

    Explanation: Lindelöf and Hausdorff do not guarantee normality without additional conditions.

119. A space with the particular point topology (all sets containing a fixed point p) is:

    (a) T_0
    (b) Not T_0
    (c) Hausdorff
    (d) Normal

    Answer: (a)

    Explanation: The particular point topology is T_0, as points can be distinguished by open sets containing p.

120. The separation axiom hierarchy is:

(a) $T_4 \implies T_3 \implies T_2 \implies T_1 \implies T_0$
(b) $T_0 \implies T_1 \implies T_2 \implies T_3 \implies T_4$
(c) $T_2 \implies T_4 \implies T_3 \implies T_1 \implies T_0$
(d) $T_1 \implies T_0 \implies T_2 \implies T_3 \implies T_4$

Answer: (a)

Explanation: The separation axioms form a hierarchy: normality (T_4) implies regularity (T_3), which implies Hausdorff (T_2), which implies T_1, which implies T_0.

Chapter 4

Connectedness and Compactness

4.1 Connected Sets and Their Properties

4.1.1 Definition of Connectedness

A set $E \subseteq \mathbb{R}$ is said to be **connected** if it cannot be partitioned into two non-empty disjoint open sets. In other words, a set is connected if there do not exist two non-empty disjoint open sets U and V such that:

$$E \subseteq U \cup V \quad \text{and} \quad E \cap U \neq \emptyset, \quad E \cap V \neq \emptyset.$$

This definition essentially means that a connected set cannot be "split" into two separate parts by open sets. Geometrically, a connected set forms a "single piece," with no gaps or separations.

Types of Connectedness

There are different types of connectedness that are often discussed in topology and analysis. For example:

- **Path-connectedness:** A set is path-connected if there exists a continuous path between any two points in the set. Path-connected sets are always connected, but not all connected sets are path-connected.

- **Connectedness in Metric Spaces:** A set is connected in a metric space if it cannot be separated by a continuous function that maps the set to a discrete space, i.e., there is no continuous function $f : X \to \{0, 1\}$ that "separates" the set.

Chapter 4: Connectedness and Compactness

Examples of Connected Sets

1. The closed interval $[0, 1]$ is connected. This is because it cannot be partitioned into two non-empty disjoint open sets within the real numbers.

2. The open interval $(0, 1)$ is connected. Similarly, it cannot be divided into two non-empty disjoint open sets in $\mathbb{R}$.

3. The entire real line $\mathbb{R}$ is connected. Any attempt to separate $\mathbb{R}$ into two non-empty disjoint open sets fails, making $\mathbb{R}$ connected.

4. Any convex subset of $\mathbb{R}^n$ is connected. For instance, any line segment or region in Euclidean space is connected.

Properties of Connected Sets

Connected sets have several important properties that play a central role in various areas of mathematics, particularly in analysis and topology.

1. The **continuous image** of a connected set is connected.
   If $f : E \to Y$ is a continuous function and $E \subseteq \mathbb{R}$ is connected, then $f(E)$ is connected in the target space Y. This is a key property used in many proofs, especially in the study of continuous functions and their behavior.

2. A connected subset of $\mathbb{R}$ is an **interval**.
   Any connected subset of $\mathbb{R}$, whether open, closed, or half-open, must be an interval. For example, the set $(1, 2) \subset \mathbb{R}$ is connected because it is an interval. In fact, the interval property is often used as a characterization of connected sets in $\mathbb{R}$.

3. **Intermediate Value Theorem:** The Intermediate Value Theorem is a fundamental result in calculus that relies on the connectedness of intervals. If $f : [a, b] \to \mathbb{R}$ is a continuous function, and y_0 is a value between $f(a)$ and $f(b)$, then there exists some $c \in [a, b]$ such that $f(c) = y_0$.
   Example: Let $f(x) = x^2$ on $[0, 2]$, where $f(0) = 0$ and $f(2) = 4$. If we take $y_0 = 1$, there exists a $c = 1$ such that $f(c) = 1$. The connectedness of the interval $[0, 2]$ ensures that this value exists.

4. If A and B are disjoint, non-empty closed sets, then the space is disconnected if there exists a separation between these sets by open sets. This property is useful when examining whether a space or set can be "broken apart" by certain types of continuous functions or mappings.

5. **Unions of Connected Sets:** The union of two connected sets A and B is connected if and only if they have a non-empty intersection. This is another key property used in constructing connected sets from simpler components.

4.1.2 Disconnected Sets

A set $E \subseteq \mathbb{R}$ is said to be **disconnected** if it can be represented as the union of two non-empty disjoint open sets. Disconnected sets are the opposite of connected sets, meaning that they can be split into separate parts, each of which is open in the relative topology. Disconnected sets often arise in analysis when we are considering spaces with certain types of boundaries or "gaps."

1. The set $\{1, 2\}$ is disconnected. It consists of two distinct points with no points in between, and hence can be separated into two distinct open sets $\{1\}$ and $\{2\}$, which form a disconnection.

2. The set $[0, 1] \cup [2, 3]$ is disconnected. Here, the two intervals are separated by the "gap" between 1 and 2, and no continuous path connects a point in $[0, 1]$ to a point in $[2, 3]$, making this set disconnected.

3. The set $\{0\} \cup (1, 2)$ is disconnected. The point $\{0\}$ is isolated, while the interval $(1, 2)$ forms a separate part, creating a disconnection.

4.1.3 Applications and Further Studies of Connectedness

The study of connected sets is vital in various branches of mathematics. Some applications include:

- In **analysis**, the connectedness of intervals is used to prove many important theorems, such as the Intermediate Value Theorem, which relies on the fact that continuous functions on intervals have the property of taking all intermediate values.

- In **differential geometry** and **manifold theory**, connectedness is essential for studying the structure of manifolds, especially when considering path-connected spaces, which are necessary for defining concepts like holonomy and geodesics.

- In **algebraic topology**, the connectedness of a space can help classify its topological properties and study its fundamental group, which captures the idea of loops and paths in the space.

- In **dynamical systems**, connectedness plays a crucial role in understanding the behavior of trajectories and attractors, ensuring that the space of possible states of a system is not fragmented into disconnected regions.

The distinction between connected and disconnected sets is crucial not only in pure mathematics but also in applied fields such as physics, engineering, and economics, where connectedness can represent continuity, stability, and the absence of abrupt changes.

Conclusion

Connectedness is a fundamental topological property that helps characterize the structure of spaces and sets. It provides insight into the behavior of functions, sequences, and spaces under continuous transformations. By understanding the properties and behavior of connected and disconnected sets, mathematicians are able to establish important results in analysis, geometry, and beyond.

4.2 Compact Sets and Compactness Criteria

4.2.1 Definition of Compactness

A subset $K \subseteq \mathbb{R}$ is said to be **compact** if every open cover of K has a finite subcover. That is, if for every collection of open sets $\{U_\alpha\}_{\alpha \in A}$ such that

$$K \subseteq \bigcup_{\alpha \in A} U_\alpha,$$

there exists a finite subcollection $\{U_{\alpha_1}, U_{\alpha_2}, \ldots, U_{\alpha_n}\}$ satisfying

$$K \subseteq \bigcup_{i=1}^{n} U_{\alpha_i}.$$

This definition ensures that no matter how we try to "cover" the set with open sets, we can always do so using only finitely many of them.

4.2.2 Characterization in $\mathbb{R}$

In the real number system, compactness can be characterized more simply using the well-known Heine–Borel Theorem:

Theorem 4.2.1 (Heine–Borel Theorem). *A set $K \subseteq \mathbb{R}$ is compact if and only if it is both **closed** and **bounded**.*

This criterion simplifies the process of checking compactness in $\mathbb{R}$, reducing it to verifying just two properties.

Examples of Compact Sets

- The closed interval $[0, 1]$ is compact because it is closed and bounded.
- Any finite set, such as $\{1\}$ or $\{1, 2, 3\}$, is compact. Finite sets are trivially covered by finitely many open sets.
- The union of finitely many compact sets is compact. For example, $[0, 1] \cup [2, 3]$ is compact since both intervals are individually compact and their union is still closed and bounded.

Examples of Non-Compact Sets

- The open interval $(0, 1)$ is not compact. While it is bounded, it is not closed, and thus fails the Heine–Borel condition.
- The set $\mathbb{R}$ is not compact because it is unbounded.
- The half-infinite interval $[0, \infty)$ is closed but unbounded, hence not compact.

4.2.3 Properties of Compact Sets

Compact sets possess several powerful and useful properties in analysis and topology:

1. **Compactness is preserved under continuous functions.**
   If $f : X \to Y$ is continuous and $K \subseteq X$ is compact, then $f(K) \subseteq Y$ is also compact.
   Example: If $f(x) = x^2$ and $K = [-2, 1]$, then $f(K) = [0, 4]$, which is compact.

Chapter 4: Connectedness and Compactness 161

2. **Closed subsets of a compact set are compact.**
   If $F \subseteq K$ is closed and K is compact, then F is also compact.
   Example: Let $K = [0, 1]$ and $F = [0, 0.5]$; then F is compact.

3. **Compact sets in $\mathbb{R}$ are bounded.**
   This is a necessary condition for compactness in real analysis. Any compact subset of $\mathbb{R}$ cannot extend infinitely in any direction.

4. **Compact sets are sequentially compact.**
   In $\mathbb{R}$, compactness is equivalent to every sequence in the set having a convergent subsequence whose limit lies within the set.

5. **Intersection property:**
   The intersection of any collection of compact sets with the finite intersection property (i.e., every finite subcollection has non-empty intersection) has a non-empty intersection.

4.2.4 Applications of Compactness

Compactness plays a central role in many theorems and applications throughout mathematics:

- In real analysis, the Extreme Value Theorem states that a continuous real-valued function on a compact set attains both a maximum and a minimum.

- In optimization problems, compactness ensures the existence of optimal solutions within feasible regions.

- In functional analysis and topology, compactness often substitutes for finiteness and is essential in the study of convergence, continuity, and completeness.

Conclusion

Compactness is a foundational concept that provides a framework for dealing with finiteness in infinite spaces. Its many characterizations and properties make it an indispensable tool in real analysis, topology, and beyond. In $\mathbb{R}$, the Heine–Borel Theorem offers a simple criterion—closed and bounded—that is both elegant and powerful for determining compactness.

4.3 Basic Concepts Related to Compactness and Connectedness

Understanding compactness and connectedness requires familiarity with several key theorems from real analysis and topology. These foundational results illustrate how continuity, boundedness, and closedness interplay in determining the behavior of functions and sequences in $\mathbb{R}$.

4.3.1 Bolzano–Weierstrass Theorem

Theorem 4.3.1. *Every bounded sequence in $\mathbb{R}$ has a convergent subsequence.*

This theorem is fundamental in real analysis and characterizes the behavior of bounded sequences, ensuring that "accumulation" points exist.

Examples

1. The sequence $x_n = (-1)^n$ is bounded but does not converge. However, the subsequence $x_{2n} = 1$ is constant and converges to 1.

2. The sequence $x_n = \frac{1}{n}$ is bounded and converges to 0, hence every subsequence also converges to 0.

3. The sequence $x_n = 1 + \frac{1}{n}$ is bounded and converges to 1. Therefore, it has convergent subsequences as well.

4.3.2 Heine–Borel Theorem

Theorem 4.3.2. *A subset $K \subseteq \mathbb{R}$ is compact if and only if it is closed and bounded.*

This result provides a practical criterion for compactness in $\mathbb{R}$, making it easy to determine whether a set is compact based solely on its structure.

Examples

1. The interval $[0, 1]$ is compact since it is both closed and bounded.

2. The interval $(0, 1)$ is not compact because it is not closed.

3. The singleton set $\{1\}$ is compact as it is finite (finite sets are always compact in $\mathbb{R}$).

Chapter 4: Connectedness and Compactness

4.3.3 Intermediate Value Theorem

Theorem 4.3.3. *Let $f : [a, b] \to \mathbb{R}$ be continuous. If $f(a) \neq f(b)$, then for every y between $f(a)$ and $f(b)$, there exists a point $c \in [a, b]$ such that $f(c) = y$.*

This theorem guarantees that continuous functions take on all intermediate values on an interval.

Examples

1. For $f(x) = x^2$ on $[0, 2]$, the function takes all values in $[0, 4]$. For example, there exists $c \in [0, 2]$ such that $f(c) = 1$.

2. For $f(x) = \sin(x)$ on $[0, \pi]$, the function assumes every value in the interval $[0, 1]$.

3. For $f(x) = x^3$ on $[-1, 1]$, the function is continuous and takes all values between -1 and 1.

4.3.4 Extreme Value Theorem

Theorem 4.3.4. *Let $f : K \to \mathbb{R}$ be a continuous function, where K is a compact subset of $\mathbb{R}$. Then f attains both its maximum and minimum values on K; that is, there exist points $x_{\min}, x_{\max} \in K$ such that*

$$f(x_{\min}) \leq f(x) \leq f(x_{\max}) \quad \text{for all } x \in K.$$

This theorem is a cornerstone of analysis and assures us that continuous functions on compact sets are bounded and reach their extrema.

Examples

1. For $f(x) = x^2$ on $[0, 2]$, the function attains its minimum at $x = 0$ and maximum at $x = 2$.

2. For $f(x) = \sin(x)$ on $[0, \pi]$, the maximum value 1 is attained at $x = \frac{\pi}{2}$, and the minimum value 0 at $x = 0$ and $x = \pi$.

3. For $f(x) = e^{-x}$ on $[0, 1]$, the maximum is $f(0) = 1$ and the minimum is $f(1) = e^{-1}$.

Conclusion

These foundational theorems interconnect ideas of boundedness, continuity, compactness, and connectedness in elegant and powerful ways. Together, they form the backbone of many analytical and topological arguments in real analysis and beyond. Recognizing these patterns enhances both intuition and problem-solving in mathematics.

4.4 Additional Concepts Related to Compactness and Connectedness

Beyond the standard theorems, several additional notions deepen our understanding of compact and connected sets. These include variations like local compactness and path connectedness, which have significant implications in analysis, topology, and applied mathematics.

4.4.1 Path Connectedness

Definition 4.4.1. *A topological space X is said to be **path connected** if for any two points $x, y \in X$, there exists a continuous function $f : [0,1] \to X$ such that $f(0) = x$ and $f(1) = y$. The function f is called a path from x to y.*

Examples

1. The interval $[0, 1]$ is path connected.

2. Any open or closed interval in $\mathbb{R}$ is path connected.

3. The space $\mathbb{R}^n$ with the Euclidean topology is path connected.

Remarks

- Every path connected space is connected, but the converse is not necessarily true.

- Example of a connected but not path connected set: The *topologist's sine curve*.

4.4.2 Locally Compact Spaces

Definition 4.4.2. *A topological space X is **locally compact** at a point $x \in X$ if there exists an open set U containing x such that the closure $\overline{U}$ is compact. If this is true for every point $x \in X$, then X is called a locally compact space.*

Chapter 4: Connectedness and Compactness 165

Examples
1. Every open subset of $\mathbb{R}^n$ is locally compact.

2. The real line $\mathbb{R}$ is locally compact.

3. The set $\mathbb{Q}$ with the usual topology is not locally compact.

4.4.3 Compactness and Sequential Compactness

Definition 4.4.3. *A topological space X is said to be **sequentially compact** if every sequence in X has a convergent subsequence whose limit is in X.*

Relation
- In metric spaces, compactness and sequential compactness are equivalent.

- However, in general topological spaces, the two notions may differ.

4.4.4 Applications of Compactness and Connectedness

1. **Existence of Maximum and Minimum Values**: By the Extreme Value Theorem, compactness ensures that continuous functions attain extrema, which is useful in optimization.

2. **Continuity and Control**: Compactness guarantees uniform continuity of functions (via the Heine–Cantor theorem).

3. **Stability of Solutions**: In differential equations and dynamical systems, compactness can guarantee the existence of solutions within bounded regions.

4. **Topological Classification**: Connectedness helps determine the number of components in a space and is central to classifying manifolds and other geometric structures.

4.4.5 Product of Compact and Connected Sets

- The product of finitely many compact spaces is compact (Tychonoff's theorem generalizes this to arbitrary products).

- The product of connected spaces is connected.

- The product of path connected spaces is path connected.

Summary of Key Topological Properties in $\mathbb{R}$

- **Compact**: A subset of $\mathbb{R}$ is compact if and only if it is closed and bounded. This is a consequence of the Heine–Borel Theorem.

- **Connected**: A set is connected if it cannot be expressed as the union of two non-empty, disjoint, open subsets. In $\mathbb{R}$, connected sets are intervals.

- **Path Connected**: A set is path connected if any two points in the set can be joined by a continuous path lying entirely within the set. In $\mathbb{R}$, all connected open sets are also path connected.

- **Sequentially Compact**: A set is sequentially compact if every sequence has a convergent subsequence whose limit lies within the set. In $\mathbb{R}$, this is equivalent to compactness.

- **Locally Compact**: A space is locally compact if every point has a neighborhood whose closure is compact. In $\mathbb{R}$, all open subsets are locally compact.

4.5 MCQs: Connectedness and Compactness

1. A set S in a topological space is said to be connected if:

    (a) S can be expressed as the union of two disjoint non-empty open sets.

    (b) S can be expressed as the intersection of two disjoint non-empty open sets.

    (c) S cannot be expressed as the union of two disjoint non-empty open sets.

    (d) S cannot be expressed as the intersection of two disjoint non-empty open sets.

    Answer: (c)
    Explanation: A set S is connected if it cannot be divided into two non-empty disjoint open sets. This means S cannot be expressed as the union of two disjoint non-empty open sets.

2. Which of the following is a criterion for compactness in a metric space?

Chapter 4: Connectedness and Compactness

(a) A set is compact if it is connected.

(b) A set is compact if every sequence in the set has a convergent subsequence whose limit is within the set.

(c) A set is compact if it is closed and bounded.

(d) A set is compact if it is open and bounded.

Answer: (b)
Explanation: In a metric space, a set is compact if and only if every sequence in the set has a convergent subsequence whose limit is within the set (this is known as the sequential compactness criterion).

3. In the context of compactness, which of the following statements is true in a general topological space?

   (a) Every compact set is closed.

   (b) Every compact set is open.

   (c) Every compact set is both open and closed.

   (d) Every compact set is closed in a Hausdorff space.

Answer: (d)
Explanation: In a Hausdorff space, every compact set is closed. However, compact sets are not necessarily open, nor do they have to be both open and closed.

4. Which of the following statements correctly describes a compact set in $\mathbb{R}^n$?

   (a) A set is compact if it is closed and bounded.

   (b) A set is compact if it is bounded.

   (c) A set is compact if it is closed.

   (d) A set is compact if it is open and bounded.

Answer: (a)
Explanation: In $\mathbb{R}^n$, a set is compact if and only if it is both closed and bounded. This is a result of the Heine-Borel theorem.

5. Which of the following is not a property of a connected space?

(a) Every continuous image of a connected space is connected.

(b) A connected space cannot be partitioned into two disjoint non-empty open subsets.

(c) Any two points in a connected space can be joined by a path.

(d) The union of two disjoint connected spaces is not connected.

Answer: (c)
Explanation: While it is true that a connected space cannot be partitioned into two disjoint non-empty open subsets, not every connected space allows for a path between any two points (unless the space is also path-connected).

6. If a set S is compact and $f : S \to \mathbb{R}$ is continuous, then $f(S)$ is:

    (a) Open in $\mathbb{R}$.

    (b) Closed in $\mathbb{R}$.

    (c) Compact in $\mathbb{R}$.

    (d) Both open and closed in $\mathbb{R}$.

Answer: (c)
Explanation: The continuous image of a compact set is compact. Since S is compact and f is continuous, $f(S)$ is compact in $\mathbb{R}$.

7. Which of the following statements about connected spaces is true?

    (a) Every connected space is path-connected.

    (b) Every path-connected space is connected.

    (c) The product of two connected spaces is disconnected.

    (d) The union of two disjoint connected spaces is connected.

Answer: (b)
Explanation: Every path-connected space is connected, but not every connected space is path-connected.

8. In a topological space, the union of a finite number of connected sets is connected if and only if:

    (a) The sets are disjoint.

Chapter 4: Connectedness and Compactness 169

(b) Each pair of sets intersects.

(c) Each pair of sets is disjoint.

(d) The sets are compact.

Answer: (b)
Explanation: The union of a finite number of connected sets is connected if and only if each pair of sets intersects.

9. Which of the following is an equivalent condition for compactness in $\mathbb{R}^n$?

    (a) A set is compact if it is closed and bounded.

    (b) A set is compact if it is open and bounded.

    (c) A set is compact if it is closed and connected.

    (d) A set is compact if every open cover has a finite subcover.

    Answer: (d)
    Explanation: In $\mathbb{R}^n$, a set is compact if and only if every open cover has a finite subcover. This is the definition of compactness.

10. A set S in a topological space is said to be compact if:

    (a) Every open cover of S has a finite subcover.

    (b) Every closed cover of S has a finite subcover.

    (c) Every open cover of S has a countable subcover.

    (d) Every closed cover of S has a countable subcover.

    Answer: (a)
    Explanation: A set S is compact if every open cover of S has a finite subcover.

11. A topological space is compact if and only if it is:

    (a) Both bounded and closed.

    (b) Both open and bounded.

    (c) Closed and every open cover has a finite subcover.

    (d) Open and every open cover has a finite subcover.

Answer: (c)
Explanation: A topological space is compact if and only if every open cover has a finite subcover.

12. Which of the following statements is true about a connected subspace of $\mathbb{R}^2$?

    (a) Every connected subspace of $\mathbb{R}^2$ is path-connected.

    (b) Every path-connected subspace of $\mathbb{R}^2$ is connected.

    (c) A connected subspace of $\mathbb{R}^2$ can be a single point or a line.

    (d) Every connected subspace of $\mathbb{R}^2$ is a curve.

    Answer: (b)
    Explanation: In $\mathbb{R}^2$, every path-connected space is connected. However, not every connected subspace is path-connected.

13. If S is a compact subset of a Hausdorff space, then S is:

    (a) Open.

    (b) Closed.

    (c) Open and closed.

    (d) Neither open nor closed.

    Answer: (b)
    Explanation: In a Hausdorff space, compact sets are closed.

14. A set S is compact in a metric space if:

    (a) It is closed.

    (b) It is bounded.

    (c) It is both closed and bounded.

    (d) It is open.

    Answer: (c)
    Explanation: In a metric space, a set is compact if and only if it is both closed and bounded.

15. The continuous image of a compact space is:

Chapter 4: Connectedness and Compactness

(a) Open.

(b) Closed.

(c) Compact.

(d) Path-connected.

Answer: (c)
Explanation: The continuous image of a compact set is compact.

16. If a space X is connected, then which of the following must be true about any subspace $Y \subseteq X$?

    (a) Y is connected.

    (b) Y is disconnected if it is not empty.

    (c) Y is connected if and only if Y is non-empty.

    (d) Y is connected if Y intersects with every connected component of X.

    Answer: (d)
    Explanation: A subspace Y of a connected space X is connected if Y intersects with every connected component of X.

17. In topology, a space is compact if every open cover has a:

    (a) Finite subcover.

    (b) Countable subcover.

    (c) Uncountable subcover.

    (d) Single open set.

    Answer: (a)
    Explanation: By definition, a space is compact if every open cover has a finite subcover.

18. The intersection of a finite number of compact sets in a topological space is:

    (a) Compact.

    (b) Open.

(c) Closed.

(d) Path-connected.

Answer: (a)

Explanation: The intersection of a finite number of compact sets is compact.

19. The union of a finite number of compact sets is:

    (a) Compact.

    (b) Open.

    (c) Closed.

    (d) Path-connected.

    Answer: (a)

    Explanation: The union of a finite number of compact sets is not necessarily compact. The property of compactness is not preserved under finite unions.

20. In a compact metric space, every sequence has:

    (a) A convergent subsequence.

    (b) A countable subsequence.

    (c) An uncountable subsequence.

    (d) A subsequence that is not bounded.

    Answer: (a)

    Explanation: In a compact metric space, every sequence has a convergent subsequence whose limit is within the space.

21. In a topological space, if a space is path-connected, then it is also:

    (a) Connected.

    (b) Compact.

    (c) Both open and closed.

    (d) A metric space.

    Answer: (a)

    Explanation: Path-connected spaces are connected.

Chapter 4: Connectedness and Compactness

22. A set S in $\mathbb{R}^2$ is connected if and only if:

    (a) It cannot be divided into two disjoint open subsets.
    (b) It is bounded.
    (c) It is closed.
    (d) It is open.

    Answer: (a)
    Explanation: A set in $\mathbb{R}^2$ is connected if it cannot be divided into two disjoint open subsets.

23. If S is compact, which of the following statements about continuous functions $f : S \to \mathbb{R}$ is true?

    (a) $f(S)$ is open in $\mathbb{R}$.
    (b) $f(S)$ is closed in $\mathbb{R}$.
    (c) $f(S)$ is compact in $\mathbb{R}$.
    (d) $f(S)$ is path-connected in $\mathbb{R}$.

    Answer: (c)
    Explanation: The continuous image of a compact set is compact.

24. In the product of two compact spaces X and Y, the product space $X \times Y$ is:

    (a) Compact.
    (b) Open.
    (c) Closed.
    (d) Path-connected.

    Answer: (a)
    Explanation: The product of two compact spaces is compact.

25. In a Hausdorff space, compact sets are:

    (a) Both open and closed.
    (b) Closed.
    (c) Open.

(d) Neither open nor closed.

Answer: (b)
Explanation: In a Hausdorff space, compact sets are closed.

26. Which of the following is not a characteristic of compact spaces?

    (a) Every open cover has a finite subcover.

    (b) Every sequence has a convergent subsequence.

    (c) The space is bounded.

    (d) The space is closed in a Hausdorff space.

Answer: (c)
Explanation: Being bounded is not a characteristic of compact spaces in general; it is only true in $\mathbb{R}^n$.

27. In a metric space, a set is compact if and only if it is:

    (a) Closed.

    (b) Open.

    (c) Closed and bounded.

    (d) Open and bounded.

Answer: (c)
Explanation: In a metric space, a set is compact if and only if it is closed and bounded.

28. Which of the following is true about the product of two connected spaces?

    (a) The product space is connected if and only if at least one of the spaces is connected.

    (b) The product space is connected if both spaces are connected.

    (c) The product space is always disconnected.

    (d) The product space is connected if and only if both spaces are disconnected.

Chapter 4: Connectedness and Compactness 175

Answer: (b)
Explanation: The product of two connected spaces is connected if and only if both spaces are connected.

29. A set S in a metric space is connected if:

    (a) It cannot be partitioned into two non-empty disjoint open sets.
    (b) It is bounded.
    (c) It is closed.
    (d) It is open.

Answer: (a)
Explanation: A set in a metric space is connected if it cannot be partitioned into two non-empty disjoint open sets.

30. In a topological space, the closure of a connected set is:

    (a) Always connected.
    (b) Always disconnected.
    (c) Open.
    (d) Bounded.

Answer: (a)
Explanation: The closure of a connected set in a topological space is always connected.

31. If S is a connected space and $A \subseteq S$, then A is connected if:

    (a) A is non-empty.
    (b) A intersects every connected component of S.
    (c) A is closed.
    (d) A is open.

Answer: (b)
Explanation: A subspace A of a connected space S is connected if A intersects every connected component of S.

32. In a topological space, a set is compact if:

(a) Every open cover has a finite subcover.

(b) Every closed cover has a finite subcover.

(c) Every open set is compact.

(d) Every closed set is compact.

Answer: (a)
Explanation: A set is compact if every open cover has a finite subcover.

33. The intersection of any number of compact sets in a topological space is:

    (a) Compact.

    (b) Disconnected.

    (c) Open.

    (d) Bounded.

    Answer: (a)
    Explanation: The intersection of any number of compact sets in a topological space is compact.

34. A set S in a Hausdorff space is compact if:

    (a) It is bounded and closed.

    (b) It is closed.

    (c) It is open and bounded.

    (d) It is closed and every open cover has a finite subcover.

    Answer: (d)
    Explanation: A set is compact in a Hausdorff space if it is closed and every open cover has a finite subcover.

35. In a topological space, which of the following is true for compact sets?

    (a) They are always open.

    (b) They are always closed.

    (c) They are always path-connected.

    (d) They are closed in a Hausdorff space.

Chapter 4: Connectedness and Compactness 177

Answer: (d)
Explanation: In a Hausdorff space, compact sets are closed.

36. A connected space that contains only one point is:

    (a) Compact.
    (b) Disconnected.
    (c) Path-connected.
    (d) Open.

    Answer: (a)
    Explanation: A space with a single point is compact as it trivially satisfies the definition of compactness.

37. The union of an arbitrary number of connected sets is connected if:

    (a) The sets are disjoint.
    (b) The sets are non-empty.
    (c) Each pair of sets intersects.
    (d) The sets are open.

    Answer: (c)
    Explanation: The union of an arbitrary number of connected sets is connected if each pair of sets intersects.

38. If S is a connected subspace of $\mathbb{R}^n$, then:

    (a) S is always path-connected.
    (b) S can be disconnected.
    (c) S is always compact.
    (d) S can be disconnected if it is not path-connected.

    Answer: (a)
    Explanation: In $\mathbb{R}^n$, connected sets are path-connected.

39. Which of the following does not necessarily imply compactness in a metric space?

    (a) Closed and bounded.

(b) Every sequence has a convergent subsequence.

(c) Every open cover has a finite subcover.

(d) Every closed subset of the space is compact.

Answer: (d)
Explanation: In general, not every closed subset of a metric space is compact.

40. A set S in a topological space is compact if every sequence in S:

    (a) Has a convergent subsequence.

    (b) Has a countable subsequence.

    (c) Is bounded.

    (d) Is finite.

    Answer: (a)
    Explanation: In a metric space, a set is compact if every sequence has a convergent subsequence whose limit is within the set.

41. The set of all limit points of a compact set is:

    (a) Compact.

    (b) Open.

    (c) Disconnected.

    (d) Path-connected.

    Answer: (a)
    Explanation: The set of all limit points of a compact set is itself compact.

42. In $\mathbb{R}^n$, the Heine-Borel theorem states that a set is compact if and only if it is:

    (a) Closed and bounded.

    (b) Open and bounded.

    (c) Closed and connected.

    (d) Open and connected.

Answer: (a)
Explanation: According to the Heine-Borel theorem, a set in $\mathbb{R}^n$ is compact if and only if it is closed and bounded.

43. In a Hausdorff space, if a set is compact and f is a continuous function from that set to another space, then:

    (a) f is open.
    (b) f is closed.
    (c) f is compact.
    (d) f is continuous.

    Answer: (d)
    Explanation: A continuous function from a compact set is continuous, but this does not imply that f is open or closed.

44. A connected space that is not path-connected:

    (a) Exists.
    (b) Cannot exist.
    (c) Is necessarily compact.
    (d) Is necessarily disconnected.

    Answer: (a)
    Explanation: It is possible for a connected space to not be path-connected.

45. In a compact Hausdorff space, every continuous function to $\mathbb{R}$ is:

    (a) Unbounded.
    (b) Not necessarily bounded.
    (c) Always bounded.
    (d) Always unbounded.

    Answer: (c)
    Explanation: In a compact Hausdorff space, every continuous function to $\mathbb{R}$ is bounded.

46. In $\mathbb{R}^n$, the intersection of two closed sets is:

(a) Open.

(b) Closed.

(c) Compact.

(d) Disconnected.

Answer: (b)
Explanation: The intersection of two closed sets in $\mathbb{R}^n$ is closed.

47. If a space S is connected, then any continuous image of S is:

    (a) Connected.

    (b) Disconnected.

    (c) Open.

    (d) Closed.

    Answer: (a)
    Explanation: The continuous image of a connected space is connected.

48. Which of the following is not necessarily true for a compact set?

    (a) It is bounded.

    (b) It is closed.

    (c) It is open.

    (d) Every sequence in it has a convergent subsequence.

    Answer: (c)
    Explanation: A compact set is not necessarily open; it is only guaranteed to be closed in a Hausdorff space.

49. The union of a finite number of compact sets is:

    (a) Compact.

    (b) Always closed.

    (c) Always bounded.

    (d) Path-connected.

    Answer: (a)
    Explanation: The union of a finite number of compact sets is compact.

Chapter 4: Connectedness and Compactness

50. In a Hausdorff space, every compact subset is:

    (a) Closed.

    (b) Open.

    (c) Both open and closed.

    (d) Path-connected.

    Answer: (a)
    Explanation: In a Hausdorff space, compact subsets are closed.

51. If X and Y are compact spaces, then the space $X \times Y$ is:

    (a) Compact.

    (b) Open.

    (c) Disconnected.

    (d) Path-connected.

    Answer: (a)
    Explanation: The product of two compact spaces is compact.

52. Which of the following is true about a connected space?

    (a) It cannot be separated into two disjoint non-empty open subsets.

    (b) It must be path-connected.

    (c) It is always compact.

    (d) It is always open in any space.

    Answer: (a)
    Explanation: A connected space cannot be separated into two disjoint non-empty open subsets.

53. In a Hausdorff space, compact subsets are:

    (a) Closed and bounded.

    (b) Always closed.

    (c) Always open.

    (d) Both open and closed.

Answer: (b)
Explanation: In a Hausdorff space, compact subsets are closed.

54. The complement of a compact set in a Hausdorff space is:

    (a) Open.
    (b) Closed.
    (c) Compact.
    (d) Disconnected.

Answer: (a)
Explanation: The complement of a compact set in a Hausdorff space is open.

55. A topological space X is compact if and only if:

    (a) Every sequence in X has a convergent subsequence.
    (b) Every open cover has a finite subcover.
    (c) Every closed cover has a finite subcover.
    (d) Every continuous function from X is bounded.

Answer: (b)
Explanation: A space is compact if and only if every open cover has a finite subcover.

56. Which of the following properties are preserved under taking subspaces?

    (a) Compactness.
    (b) Connectedness.
    (c) Both compactness and connectedness.
    (d) Path-connectedness.

Answer: (b)
Explanation: Connectedness is preserved under taking subspaces, but compactness is not necessarily preserved.

57. In a topological space, if every sequence has a convergent subsequence, then the space is:

(a) Compact.

(b) Connected.

(c) Hausdorff.

(d) Path-connected.

Answer: (a)
Explanation: In a metric space, if every sequence has a convergent subsequence, then the space is compact.

58. A space X is compact if every continuous function $f : X \to \mathbb{R}$ is:

    (a) Unbounded.

    (b) Bounded.

    (c) Continuous.

    (d) Path-connected.

Answer: (b)
Explanation: In a compact space, every continuous function to $\mathbb{R}$ is bounded.

59. The product of two connected spaces is:

    (a) Connected.

    (b) Disconnected.

    (c) Open.

    (d) Closed.

Answer: (a)
Explanation: The product of two connected spaces is connected.

60. If S is a connected subset of $\mathbb{R}^n$, then:

    (a) It must be path-connected.

    (b) It must be compact.

    (c) It must be closed.

    (d) It must be bounded.

Answer: (a)
Explanation: In $\mathbb{R}^n$, connected subsets are path-connected.

61. A set is compact if it is:

    (a) Closed and bounded.

    (b) Open and bounded.

    (c) Closed and connected.

    (d) Open and connected.

    Answer: (a)
    Explanation: In $\mathbb{R}^n$, a set is compact if and only if it is closed and bounded.

62. A continuous image of a compact set is:

    (a) Compact.

    (b) Open.

    (c) Closed.

    (d) Path-connected.

    Answer: (a)
    Explanation: The continuous image of a compact set is compact.

63. In $\mathbb{R}^n$, if a set is closed and bounded, then:

    (a) It is compact.

    (b) It is open.

    (c) It is connected.

    (d) It is path-connected.

    Answer: (a)
    Explanation: According to the Heine-Borel theorem, in $\mathbb{R}^n$, a set that is closed and bounded is compact.

64. In a Hausdorff space, compact sets are:

    (a) Always closed.

Chapter 4: Connectedness and Compactness

(b) Always open.

(c) Both open and closed.

(d) Always path-connected.

Answer: (a)
Explanation: In a Hausdorff space, compact sets are closed.

65. The union of two compact sets is:

    (a) Compact if the intersection of the two sets is non-empty.

    (b) Compact if the sets are disjoint.

    (c) Compact if one of the sets is bounded.

    (d) Always compact.

Answer: (a)
Explanation: The union of two compact sets is compact if their intersection is non-empty.

66. In a topological space, which of the following is true about the union of a finite number of connected sets?

    (a) It is connected if the sets have a common point.

    (b) It is connected if the sets are disjoint.

    (c) It is connected if each set is compact.

    (d) It is connected if the sets are open.

Answer: (a)
Explanation: The union of a finite number of connected sets is connected if the sets have at least one common point.

67. In a compact metric space, every continuous function to $\mathbb{R}$ is:

    (a) Unbounded.

    (b) Not necessarily bounded.

    (c) Always bounded.

    (d) Always unbounded.

Answer: (c)
Explanation: In a compact metric space, every continuous function to $\mathbb{R}$ is always bounded.

68. A space is compact if:

    (a) Every open cover has a finite subcover.
    (b) Every closed cover has a finite subcover.
    (c) Every sequence has a convergent subsequence.
    (d) Every continuous function is bounded.

    Answer: (a)
    Explanation: A space is compact if and only if every open cover has a finite subcover.

69. In a Hausdorff space, if a set is compact and f is a continuous function from that set to another space, then f is:

    (a) Open.
    (b) Closed.
    (c) Compact.
    (d) Continuous.

    Answer: (d)
    Explanation: A continuous function from a compact set is continuous.

70. If a space is compact and Hausdorff, then:

    (a) Every subset is closed.
    (b) Every continuous function to $\mathbb{R}$ is bounded.
    (c) Every compact subset is open.
    (d) Every sequence has a convergent subsequence.

    Answer: (b)
    Explanation: In a compact Hausdorff space, every continuous function to $\mathbb{R}$ is bounded.

71. In a Hausdorff space, a set is compact if:

(a) It is closed.

(b) It is bounded.

(c) Every sequence in the set has a convergent subsequence.

(d) Every open cover has a finite subcover.

Answer: (d)
Explanation: In a Hausdorff space, a set is compact if every open cover has a finite subcover.

72. The continuous image of a compact set is:

    (a) Compact.

    (b) Open.

    (c) Closed.

    (d) Connected.

    Answer: (a)
    Explanation: The continuous image of a compact set is compact.

73. The product of two compact spaces is:

    (a) Compact.

    (b) Connected.

    (c) Disconnected.

    (d) Open.

    Answer: (a)
    Explanation: The product of two compact spaces is compact.

74. If X is a connected space and $f : X \to \mathbb{R}$ is a continuous function, then $f(X)$ is:

    (a) Connected.

    (b) Disconnected.

    (c) Open.

    (d) Closed.

Answer: (a)
Explanation: The continuous image of a connected space is connected.

75. In a compact metric space, every sequence has a:

    (a) Convergent subsequence.
    (b) Countable subsequence.
    (c) Unbounded subsequence.
    (d) Bounded subsequence.

Answer: (a)
Explanation: In a compact metric space, every sequence has a convergent subsequence.

76. A connected space that is not compact:

    (a) Can exist.
    (b) Cannot exist.
    (c) Is necessarily bounded.
    (d) Is necessarily disconnected.

Answer: (a)
Explanation: It is possible for a space to be connected but not compact.

77. In a Hausdorff space, compact subsets are:

    (a) Always closed.
    (b) Always open.
    (c) Both open and closed.
    (d) Always path-connected.

Answer: (a)
Explanation: In a Hausdorff space, compact subsets are closed.

78. The intersection of an arbitrary number of compact sets is:

    (a) Compact.
    (b) Disconnected.

Chapter 4: Connectedness and Compactness

(c) Open.

(d) Bounded.

Answer: (a)

Explanation: The intersection of any number of compact sets is compact.

79. If a space is compact and Hausdorff, then:

    (a) Every subset is closed.

    (b) Every continuous function to $\mathbb{R}$ is bounded.

    (c) Every compact subset is open.

    (d) Every sequence has a convergent subsequence.

    Answer: (b)

    Explanation: In a compact Hausdorff space, every continuous function to $\mathbb{R}$ is bounded.

80. A set S in a topological space is compact if every sequence in S:

    (a) Has a convergent subsequence.

    (b) Has a countable subsequence.

    (c) Is bounded.

    (d) Is finite.

    Answer: (a)

    Explanation: In a metric space, a set is compact if every sequence has a convergent subsequence.

81. If a space is connected and Hausdorff, then:

    (a) It is compact.

    (b) It is path-connected.

    (c) Every continuous function from it to $\mathbb{R}$ is bounded.

    (d) It cannot be separated into two disjoint non-empty open subsets.

    Answer: (d)

    Explanation: In a Hausdorff space, a connected space cannot be separated into two disjoint non-empty open subsets.

82. A topological space X is connected if:

    (a) It cannot be written as the union of two disjoint non-empty open sets
    (b) It is compact
    (c) It is Hausdorff
    (d) It is finite

    Answer: (a)
    Explanation: A space is connected if it has no non-trivial clopen sets, meaning it cannot be split into two disjoint non-empty open sets.

83. The real line $\mathbb{R}$ with the standard topology is:

    (a) Connected
    (b) Not connected
    (c) Compact
    (d) Finite

    Answer: (a)
    Explanation: $\mathbb{R}$ is connected, as intervals are the only connected subsets of $\mathbb{R}$.

84. A space is path-connected if:

    (a) Any two points can be joined by a continuous path
    (b) It is connected
    (c) It is compact
    (d) It is Hausdorff

    Answer: (a)
    Explanation: Path-connectedness means there exists a continuous function from $[0, 1]$ joining any two points.

85. Every path-connected space is:

    (a) Connected
    (b) Not necessarily connected

Chapter 4: Connectedness and Compactness

(c) Compact

(d) Hausdorff

Answer: (a)

Explanation: Path-connectedness implies connectedness, as a disconnection would prevent continuous paths between points.

86. The interval $[0, 1] \subset \mathbb{R}$ with the standard topology is:

    (a) Path-connected

    (b) Not path-connected

    (c) Not connected

    (d) Open

    Answer: (a)

    Explanation: $[0, 1]$ is path-connected, as any two points can be joined by a straight-line path.

87. The image of a connected set under a continuous function is:

    (a) Connected

    (b) Compact

    (c) Open

    (d) Closed

    Answer: (a)

    Explanation: Continuous functions preserve connectedness, as a disconnection in the image would imply one in the domain.

88. A space is totally disconnected if:

    (a) Its only connected subsets are singletons

    (b) It is not connected

    (c) It is compact

    (d) It is Hausdorff

    Answer: (a)

    Explanation: A totally disconnected space has no non-trivial connected subsets.

89. The rational numbers $\mathbb{Q} \subset \mathbb{R}$ with the subspace topology are:

    (a) Totally disconnected

    (b) Connected

    (c) Compact

    (d) Path-connected

    Answer: (a)
    Explanation: $\mathbb{Q}$ is totally disconnected, as any subset with more than one point can be split by irrational numbers.

90. A space is locally connected if:

    (a) Every point has a connected neighborhood

    (b) The space is connected

    (c) The space is compact

    (d) The space is Hausdorff

    Answer: (a)
    Explanation: Local connectedness means each point has a basis of connected open sets.

91. The real line $\mathbb{R}$ with the standard topology is:

    (a) Locally connected

    (b) Not locally connected

    (c) Compact

    (d) Finite

    Answer: (a)
    Explanation: Every point in $\mathbb{R}$ has connected neighborhoods (intervals), so it is locally connected.

92. A subset $K \subseteq X$ is compact if:

    (a) Every open cover of K has a finite subcover

    (b) It is closed

    (c) It is bounded

Chapter 4: Connectedness and Compactness

(d) It is finite

Answer: (a)
Explanation: Compactness is defined by the finite subcover property for open covers.

93. The interval $[0, 1] \subset \mathbb{R}$ with the standard topology is:

    (a) Compact
    (b) Not compact
    (c) Open
    (d) Infinite

Answer: (a)
Explanation: By the Heine-Borel theorem, $[0, 1]$ is compact as it is closed and bounded in $\mathbb{R}$.

94. In a Hausdorff space, every compact subset is:

    (a) Closed
    (b) Open
    (c) Dense
    (d) Finite

Answer: (a)
Explanation: Compact subsets in Hausdorff spaces are closed, as points outside can be separated from the subset.

95. The image of a compact set under a continuous function is:

    (a) Compact
    (b) Open
    (c) Closed
    (d) Dense

Answer: (a)
Explanation: Continuous functions preserve compactness, as open covers of the image lift to the domain.

96. A space is sequentially compact if:

    (a) Every sequence has a convergent subsequence
    (b) Every sequence converges
    (c) The space is compact
    (d) The space is connected

    Answer: (a)
    Explanation: Sequential compactness means every sequence has a subsequence converging to a point in the space.

97. In a metric space, a subset is compact if and only if it is:

    (a) Closed and bounded
    (b) Open and bounded
    (c) Closed and unbounded
    (d) Open and unbounded

    Answer: (a)
    Explanation: The Heine-Borel theorem states that in $\mathbb{R}^n$, compact sets are closed and bounded.

98. The space $\mathbb{R}$ with the standard topology is:

    (a) Not compact
    (b) Compact
    (c) Closed
    (d) Finite

    Answer: (a)
    Explanation: $\mathbb{R}$ is not compact, as it is unbounded and not covered by finitely many bounded open sets.

99. A space is limit point compact if:

    (a) Every infinite subset has a limit point
    (b) Every sequence converges
    (c) The space is compact

(d) The space is connected

Answer: (a)
Explanation: Limit point compactness means every infinite set has at least one limit point.

100. In a first-countable space, compactness is equivalent to:

    (a) Sequential compactness

    (b) Connectedness

    (c) Local compactness

    (d) Separability

Answer: (a)
Explanation: In first-countable spaces, compactness, sequential compactness, and limit point compactness are equivalent.

101. Tychonoff's theorem states that:

    (a) The product of any collection of compact spaces is compact

    (b) The product of connected spaces is connected

    (c) The product of Hausdorff spaces is Hausdorff

    (d) The product of separable spaces is separable

Answer: (a)
Explanation: Tychonoff's theorem guarantees that arbitrary products of compact spaces are compact in the product topology.

102. The product $[0, 1] \times [0, 1] \subset \mathbb{R}^2$ is:

    (a) Compact

    (b) Not compact

    (c) Open

    (d) Infinite

Answer: (a)
Explanation: By Tychonoff's theorem, the product of compact spaces $[0, 1]$ is compact.

103. A space is locally compact if:

    (a) Every point has a compact neighborhood
    (b) The space is compact
    (c) The space is connected
    (d) The space is finite

    Answer: (a)
    Explanation: Local compactness means every point is contained in a compact set's interior.

104. The space $\mathbb{R}$ with the standard topology is:

    (a) Locally compact
    (b) Not locally compact
    (c) Compact
    (d) Finite

    Answer: (a)
    Explanation: Every point in $\mathbb{R}$ has a compact neighborhood, e.g., a closed interval.

105. The Cantor set is:

    (a) Compact
    (b) Not compact
    (c) Open
    (d) Connected

    Answer: (a)
    Explanation: The Cantor set is closed and bounded in $\mathbb{R}$, hence compact by the Heine-Borel theorem.

106. A closed subset of a compact space is:

    (a) Compact
    (b) Open
    (c) Connected

Chapter 4: Connectedness and Compactness

(d) Dense

Answer: (a)
Explanation: Closed subsets inherit compactness, as open covers of the subset extend to the space.

107. The product of two connected spaces is:

    (a) Connected

    (b) Not necessarily connected

    (c) Compact

    (d) Hausdorff

    Answer: (a)
    Explanation: The product of connected spaces is connected, as it cannot be split into disjoint open sets.

108. The space $\mathbb{Q} \subset \mathbb{R}$ with the subspace topology is:

    (a) Not connected

    (b) Connected

    (c) Compact

    (d) Closed

    Answer: (a)
    Explanation: $\mathbb{Q}$ is totally disconnected, as it can be split by irrational numbers.

109. A compact subset of a Hausdorff space is:

    (a) Closed

    (b) Open

    (c) Connected

    (d) Dense

    Answer: (a)
    Explanation: In Hausdorff spaces, compact subsets are closed due to separation properties.

110. The one-point compactification of $\mathbb{R}$ is homeomorphic to:

    (a) The circle S^1

    (b) $\mathbb{R}$

    (c) $[0, 1]$

    (d) $\mathbb{R}^2$

    Answer: (a)
    Explanation: Adding a point at infinity to $\mathbb{R}$ forms a space homeomorphic to the circle S^1.

111. A space is connected if and only if:

    (a) It has no non-trivial clopen sets

    (b) It is compact

    (c) It is Hausdorff

    (d) It is finite

    Answer: (a)
    Explanation: A space is connected if the only clopen sets are $\emptyset$ and X.

112. The unit circle $S^1 \subset \mathbb{R}^2$ is:

    (a) Connected

    (b) Not connected

    (c) Not compact

    (d) Open

    Answer: (a)
    Explanation: S^1 is connected, as it is path-connected (any two points can be joined by an arc).

113. A space is compact if and only if:

    (a) Every open cover has a finite subcover

    (b) It is closed

    (c) It is bounded

(d) It is connected

Answer: (a)
Explanation: This is the definition of compactness in a topological space.

114. The discrete topology on any set is:

    (a) Compact if finite

    (b) Always compact

    (c) Never compact

    (d) Connected

Answer: (a)
Explanation: In the discrete topology, a space is compact only if it is finite, as infinite sets have covers with no finite subcover.

115. A continuous function from a compact space to $\mathbb{R}$:

    (a) Is bounded

    (b) Is unbounded

    (c) Is open

    (d) Is bijective

Answer: (a)
Explanation: The image of a compact set is compact in $\mathbb{R}$, hence closed and bounded.

116. The product $\mathbb{R}^\omega$ (countable product of $\mathbb{R}$) is:

    (a) Not compact

    (b) Compact

    (c) Closed

    (d) Finite

Answer: (a)
Explanation: $\mathbb{R}$ is not compact, so the infinite product is not compact by Tychonoff's theorem.

117. A space is locally path-connected if:

    (a) Every point has a path-connected neighborhood

    (b) The space is path-connected

    (c) The space is compact

    (d) The space is Hausdorff

    Answer: (a)
    Explanation: Local path-connectedness means each point has a basis of path-connected open sets.

118. The topologist's sine curve is:

    (a) Connected but not path-connected

    (b) Path-connected

    (c) Not connected

    (d) Compact

    Answer: (a)
    Explanation: The topologist's sine curve is connected but not path-connected, as paths cannot reach the y-axis.

119. A compact space is:

    (a) Normal if Hausdorff

    (b) Always normal

    (c) Always connected

    (d) Always separable

    Answer: (a)
    Explanation: Compact Hausdorff spaces are normal, as they allow separation of closed sets.

120. The product $\prod_{n=1}^{\infty} [0, 1]$ is:

    (a) Compact

    (b) Not compact

    (c) Open

(d) Connected

Answer: (a)
Explanation: By Tychonoff's theorem, the countable product of compact spaces $[0, 1]$ is compact.

121. A subspace $Y \subseteq X$ is compact if:

    (a) It is closed and X is compact
    (b) It is open
    (c) It is dense
    (d) It is infinite

    Answer: (a)
    Explanation: A closed subset of a compact space is compact.

122. The subspace $(0, 1) \subset \mathbb{R}$ is:

    (a) Not compact
    (b) Compact
    (c) Closed
    (d) Finite

    Answer: (a)
    Explanation: $(0, 1)$ is not compact, as it is not closed in $\mathbb{R}$.

123. A continuous bijection from a compact space to a Hausdorff space is:

    (a) A homeomorphism
    (b) Not a homeomorphism
    (c) Open but not continuous
    (d) Surjective but not injective

    Answer: (a)
    Explanation: Such a map is closed, hence a homeomorphism, as compact sets map to closed sets in Hausdorff spaces.

124. The Sorgenfrey line (lower limit topology on $\mathbb{R}$) is:

(a) Not compact

(b) Compact

(c) Closed

(d) Finite

Answer: (a)
Explanation: The Sorgenfrey line is not compact, as it is not bounded.

125. A space is connected if:

    (a) Every continuous function to $\{0, 1\}$ (discrete) is constant

    (b) It is compact

    (c) It is Hausdorff

    (d) It is finite

    Answer: (a)
    Explanation: Connectedness means the only continuous maps to a discrete space are constant.

126. The co-finite topology on an infinite set is:

    (a) Compact

    (b) Not compact

    (c) Connected

    (d) Path-connected

    Answer: (a)
    Explanation: The co-finite topology is compact, as any open cover has a finite subcover due to finite complements.

127. A space is compact if:

    (a) Every filter has a cluster point

    (b) It is connected

    (c) It is Hausdorff

    (d) It is separable

Answer: (a)

Explanation: Compactness can be characterized by the convergence of filters.

128. The subspace $\mathbb{Z} \subset \mathbb{R}$ with the subspace topology is:

    (a) Not connected

    (b) Connected

    (c) Compact

    (d) Closed

    Answer: (a)

    Explanation: $\mathbb{Z}$ is disconnected, as singletons are open in the subspace topology.

129. A locally compact Hausdorff space has a one-point compactification that is:

    (a) Hausdorff

    (b) Not Hausdorff

    (c) Not compact

    (d) Connected

    Answer: (a)

    Explanation: The one-point compactification of a locally compact Hausdorff space is compact and Hausdorff.

www.ingramcontent.com/pod-product-compliance
Lightning Source LLC
Chambersburg PA
CBHW052251220526
45471CB00001B/289